ALSO BY FERGAL NALLY

*/ Non Fiction*
A History of Ashford Castle
A Manual of Oral Medicine
*/ Fiction*
A Matter of Conscience
A Matter of Time

Born in Dublin Fergal Nally is a doctor of Medicine, M.D. For many years he was Head of a clinical department in the University of London and Consultant in several London hospitals. Pain and cancer research formed a large part of his work and resulted in many publications in the scientific and medical literatures. He is also a Fellow of the Royal Society of Arts.

Order this book online at www.trafford.com
or email orders@trafford.com

Most Trafford titles are also available at major online book retailers.



Printed in Victoria, BC, Canada.

ISBN: 978-1-4269-2121-6 (sc)

ISBN: 978-1-4269-2122-3 (dj)

Library of Congress Control Number: 2009914033

Cover Design / Artwork Fergal Nally
Courtesy of Combridges Fine Arts Ltd. Dublin 2 www.cfa.ie

*Our mission is to efficiently provide the world's finest, most comprehensive book publishing service, enabling every author to experience success. To find out how to publish your book, your way, and have it available worldwide, visit us online at www.trafford.com*

*Trafford rev. 3/1/2010*

www.trafford.com

**North America & international**
toll-free: 1 888 232 4444 (USA & Canada)
phone: 250 383 6864 • fax: 812 355 4082

For Anne

# TO SPEAK
# THE UNSPOKEN

# Part One

# 1

Karen Smith lies in a peaceful haze, similar to a lethargy that comes over one on a summer's day, when sultriness persuades birds and insects into a natural siesta. She relaxes on the edge of a meadow full of wild flowers and sees light shimmering through oak leaves beyond which drifts an unending blue world. Occasionally, an arrow of pain shoots through her like lightning but the intervals become fewer and short-lived. The voices around her are far away and she pays little attention. She feels herself sinking deeper and offers little resistance to an enveloping numbness; there is no desire or effort for reattachment to the dissolving edge of consciousness.

Previously, opposition and struggle made their presence felt, but now they are no more. Fleeting images of arrogance and torment stress and worry race through her mind. Then there moves primal sensations of termination, of ending it all and arriving.

Images grow tired and fade into a thickening veil embracing her, darkness replaces the blueness. Slowly she sinks into it with serenity and without fear. It's like a rising tide gradually engulfing her, moving higher and higher – now submerging her body and creeping inexorably over her chest, throat, and mouth. Now she's smothering. Silence.

'It's finished.' Sister Mortimer closes the eyes gently and turns to the others. 'We'll have to find the husband as soon as possible.'

After Karen's death Peter Smith's attitude changes from injured husband suffering loss and disbelief, to the start of an unhealthy revenge – a revenge that is ill-defined and something that is difficult to explain. Yet it is there and begins to grow. A strategy has to be worked out.

Time is a blur after he leaves Addenbrooke's Hospital and returns to Welwyn. Several chaotic days elapse before the funeral. Most of Karen's relations live locally and busy themselves with arrangements for the event. He cannot recall what he did during those fateful days and because he has no close family the house is taken over by friends and neighbours, quietly spoken women, some he'd never seen before and men who look like a cross between civil servants and polite waiters who speak in soft undertones about technical matters – most times giving answers to their own questions. Someone asks him what is to be engraved on the upper plate of the coffin and he merely replies: 'Just her name.'

He realises he should be more explicit but people are understanding and put it down to his being overburdened. Everyone is sympathetic, it is almost suffocating.

There is one incident that causes some consternation. A well-meaning motherly person approaches and asks him in obsequious tones.

'Would you like to see her for the last time?'

He's taken aback and doesn't reply immediately.

The poor lady looks surprised and shocked.

'They're about to close the coffin and you haven't said goodbye yet. This is the last opportunity.'

Oh, it doesn't matter – he's about to say but holds back. Just to please others he goes upstairs to have a look at her. And that is that. Everyone is relieved and shows approval.

*

Up to now Dr. John Nicholson's career has been devoted to the fight against cancer. Everyday, week and month sees progress in the grim battle against the monstrous killer of young and old. Cambridge University is a wonderful centre for research and has great advantages on offer. With them come sacrifices and betrayals which he does his best to ignore. For years he strives for greater success not knowing his efforts contain the seeds of a growing tragedy. He is now in danger of losing the fight against certain forces and pays the penalty for committing that unforgivable sin – success in the scientific world. He cannot expend so much effort without a price, a feeling of being used.

Months later a phone rings sharply throughout the house -- his soul sinks from fear of further hurt or incalculable change. He sits rigidly as it rings again and again. He dreads the small black instrument shrilling and penetrating the room. It pauses for a moment, replaced by his heartbeat, and then it starts again.

There can be few things more terrifying than a phone call in the middle of the night -- at least for some especially Dr. Nicholson. Normally it wouldn't matter; it would be part of his work. Now it's different and an old terror rises in him.

There are probably times in life when we experience fear, horror, panic – call them what you will. When everything you've worked for, slaved over, given your best is to no avail. Just wasted. Worse still is the threat of punishment for your efforts. Questions like... if only...why bother...loom ahead. Friends tried to reassure him, saying things would turn out all right. He must understand. But he could understand and that was the trouble.

He lifts the phone. The case is set to go to trial in two weeks time in court.

Replacing the receiver he dresses and walks down the steps of his country cottage into the obliterating moonlight. In a moment he becomes a shadow in the street lamps and then a soft footfall on the road, as he consoles himself with the thought that conflict may have a certain advantage over victory and defeat and those long years of effort he would have forever.

No matter what lies ahead.

Two weeks later, as dawn awakens across the flat landscape flinging shadows into green and purple spaces John Nicholson is in a dreadful state. He's been driven to court to face charges of unlawful killing by administering unlicensed agents to a cancer patient. Even though, an excellent Queen's Council has agreed to act for the defence he's worried. Very worried.

'Stop upsetting yourself, John,' Karita Isselherg reassures him as she grips the steering wheel. 'It will probably turn out in your favour.'

Little does she realize.

'Something will probably come of it if I know Peter Smith.' John winces at the name. 'It's wrong to think that most people are satisfied with peace; they must have action, and will cause it if they cannot find it. People are never more dangerous when they are convinced they are right. He's dangerous and has a hidden agenda. There's a vendetta against me over the death of his wife. Karita, do you realise if the case goes against me I could end up in jail?'

'Oh John. No.' She turns to him confused. 'Whatever happens we'll fight.'

Even though their engagement has been a strain how wonderful it is to have her with him, confident intelligent and hopeful.

'You always have,' he says slowly.

'And we've chosen the best lawyers.'

'Will you stay with me in court?'

'Of course.'

'I'd like you to be there. Maybe with the right lawyers we can produce a miracle,' he says. 'We started with one. This could be another.'

He closes his eyes as they travel and remembers the first miracle.

A long time ago.

# 2

Welwyn Garden City had a well-ordered appearance in the nineteen-eighties. Roads were wide and tree-lined and houses, similar in appearance, had a geometric neatness. The place was unified; a model of what a modern city should be with parks and shops within reach of residents. Even the industrial estates were discrete and well planned.

James Wilkinson was born in a house on the south side of the city and there he'd been happy. He'd learned about caring and compassion, which he carried into later life and discovered those hidden qualities, rare in the chaos of modern living. They had their nourishment here along with experiencing the hardships of poverty.

It seemed a short time since childhood, running and laughing in this place, yet from the onset the pattern of life was directed towards the welfare of others. This sense stayed and whispered he was going to be a successful doctor. Eventually he became Head of Oncology at the nearby modern Q E 2 Hospital.

Trees filtered sun onto sparkling grass standing like elegant ladies with feathered fans flirting boldly with red roofed buildings as he walked to work. Others were bowing and fawning on flowerbeds and pathways and more spread into the distance over fields and hills making the city a lovely blend of countryside and town land.

His thoughts turned to a new patient – a woman with anxious eyes, golden hair and of an age that trembles between the abiding freshness of young life and the insistent call of forty years; a rare beauty in her face that youth sometimes gives just before it departs. She'd developed a lymphoma – a dangerous form of cancer of the lymph glands. She was on maximum chemotherapy and responded well -- for a while. A few days ago she took a turn for the worse. He presented her case at the clinico-pathological conference.

The meeting started with routine problems. Towards the end Dr. Wilkinson caught the chairman's eye.

'Thank you, sir,' he said, 'I'd like the views of colleagues on a difficult patient, a Mrs. Sylvia Clarke -- a lovely but tragic woman with acute lymphoma. She's been responding to maximum chemotherapy since admission, but in the last three days deterioration has set in. Either the drugs are losing efficacy or the condition is getting out of control. I've reached a stalemate.'

'You're sure you've reached maximum dosage?' Someone asked.

'Certainly have. Anything further would be dangerous.'

'How about combined radiotherapy?'

'She's been through that in the past and had severe reactions. If we went back now while on maximum therapy there's a chance we would kill her.'

Silence for a while. Dr. Gordon Sinclair, a recently appointed registrar who'd worked in Addenbrooke's in Cambridge mentioned Dr. Nicholson's work. He was present when Nicholson treated Annette Nolan successfully for leukaemia by marrow implants and in vitro agents. The meeting was adjourned.

Wilkinson telephoned Cambridge that afternoon.

'May I speak to Dr. John Nicholson?' He asked the secretary. 'My name is Dr. James Wilkinson, Oncologist in the Q E 2 Hospital in Welwyn Garden City.'

'I'll see if he's free.'

'Dr. Wilkinson,' a soft voice came on the line, 'Nicholson here. How can I help?'

'Good of you to take my call doctor. I appreciate it. The reason I'm phoning is we've a difficult management problem and Dr. Sinclair suggested I speak to you.'

'I know Graham. Pleasant guy and excellent doctor,' Nicholson said simply.

'He speaks well of his time at Addenbrooke's. But to come to the point a young lady with lymphoma is not responding to chemotherapy. In fact she's deteriorating by the day and our hands are tied.'

'We've had similar problems here.' John replied guardedly.

'Dr. Nicholson, would you mind if I drove to Cambridge to discuss the case with you? I've reached the end of the road and the phone is unsatisfactory. I could take the case records with me.'

'That would be a help,' John agreed. 'Why not come along as soon as you can.'

James' face brightened immediately. 'Tomorrow afternoon at 2 p.m. I could be there.'

'Excellent. I'll see you then. Come straight in the main entrance and park in the staff car-park. I'll let them know you're coming. Ask for me at Reception.'

'Wonderful. Looking forward to seeing you.'

As James returned home later that evening the courteous trees grew garrulous overhead and birds took up the burden of song in the dusky air. Before reaching home he noticed a single man sitting on a park bench, guarding an old bag of belongings and a dog at his feet. The face, the colour of bronze, was staring across the way at the brightly lit houses where a different race of people lived who still had hope.

It was a fine spring morning as James Wilkinson drove north on the A 1 M past Stevenage, turned off at the sign for Baldock. At the top of the town he joined the A 505 through rolling countryside and entered the M 11 north to Cambridge. He'd stopped at Royston for lunch.

Reaching Addenbrooke's he found a parking space. Arriving a half-hour early he'd an opportunity to examine the famous hospital. Walking around the grounds he explored as much of the exterior as possible. It was a fine, well-proportioned building and was in harmony with the city -- a feeling of conformity about things.

Time to go to Reception. The lady at the desk phoned Oncology. Dr. Nicholson would be down right away.

'Dr. Wilkinson. Welcome to Addenbrooke's.' John appeared suddenly extending a firm hand. 'Have you been here before?'

'I haven't. It's striking and, of course, has a wonderful reputation.'

'That's another day's story. Let's go to my department and discuss the problem.'

'Yes indeed. How remiss of me. It must be lovely to work and to live here.'

As John led the way a smile lightened his calm classic face.

'It is. Although there are certain drawbacks.'

*

Meanwhile in a nightclub in Hastings Peter Smith thought he was enjoying himself, having a great time and catching up on all the things he'd missed. His new girl for the night, Penny, had seen more of reality and courage, terror and pain than most of the pompous wallflowers who frequented such a world. And Peter was almost twice her age.

However, as she was young this artificial world was awash with flowers, bright and clever snobbery and music that set the mood of the crowd and summed up the seductiveness of the times. Why not make the most of it? All night the guttural saxophones, the screeching guitars and booming basses throbbed and wailed the glad new rhythms with unstoppable fever. As time wore on she blossomed more into the thrill of things – everywhere the club pulsated incessantly with a kind of fever while fresh young faces raced here and there like rose petals blown in the wind.

He looked up at her as she approached and she felt a sudden shock. He was worn to a shadow and his eyes, sunken in darkness, were smouldering in red rims of fire.

'Let's go back to the hotel. I've had enough excitement for one night,' he suggested.

For a moment the music descended to a menacing minor key and reminded him of his ongoing domestic problems – a sickly complaining wife who found it hard to give him back the custody of his leisure hours. Tonight he was here under false pretences; he was supposed to be at a week long conference of the Association of Investigative Journalists in Bournemouth.

# 3

Dr. Wilkinson was shown into the hospital Meetings' Room with its elegant table and chairs. John asked him to be seated while he was called to the phone.

Wilkinson -- ever curious -- noticed antique furniture and cabinets showing trophies of past staff. Portraits of distinguished dons and professors gazed from walls, along with two landscapes with ornate guilt frames and lit by overhead lighting on the wall behind which a chairman would sit. Close up they were impressive; he wanted to know more. In the lower corner was a signature but no indication what the scenes were although one depicted an ancient burial site surrounded by hills in the evening sun; the other a strange flat-topped mountain and a beach in the foreground. It reminded him of childhood holidays full of new sights, sounds and fragrances. Happy days that felt as long as ten days now. Nicholson interrupted his thoughts.

'Dr. Nicholson these paintings are exquisite especially hanging here.'

'Glad you like them,' John replied with a smile, 'they're unusual but seem to fit in well if we judge from comments.'

'What's the history -- if I may ask?'

'Course you may,' John replied, the words sounding like a

sigh. 'Nolan's daughter, Annette, was treated here recently for acute leukaemia by marrow transplantation. She'd little success elsewhere and was deteriorating before she arrived. It was a traumatic experience and the parents had a difficult time. She pulled through and is in remission for nearly a year. The father was so pleased he presented the pictures to the hospital and one to me.'

'They're beautiful. He must be successful.'

'He's a fair reputation in his native Dublin with sell-out exhibitions and is also quite a character.'

John gestured to a chair inviting Wilkinson to be seated.

'What can I do for you? We could discuss little on the phone.'

'Dr. Nicholson. My problem ...'

John raised a hand. 'Name is John. Please.'

'Fine. Mine's James. Well, my patient may be similar to Annette Nolan. She was admitted two weeks ago to the Q E 2 with an aggressive lymphoma. And try as we may with every combination she showed no response. She's going downhill and there's nothing we can do. We're helpless and yet we don't want to give up.'

Footfalls of memory echoed in John's mind. 'I see. I've been there before.'

'Dr. Sinclair mentioned your work. All her records and reports are with me. They're copies and you're welcome to keep them.'

John looked away trying to find the best way forward. This could be difficult. A patient should be left in the same hospital to recover or die – but his way of thinking was not conventional. Perhaps there was another option – his experimental way. But this would take courage -- a lot of courage and sometimes courage may only be found on considered reflection.

After a few moments he said slowly. 'There are two options. Firstly, she could be transferred to Cambridge and we could take over.'

James nodded yet was doubtful.

'The second would be for her to stay at the Q E 2. You could start the in vitro treatment we're using in our double blind trial -- with the exception she's been on conventional treatment and we know the result. Let's see the response to new therapy in the same setting.'

James' face lit up. 'The second would be better. She's familiar with us and under our care for months. Her family and friends live close and are a wonderful support. A good test of your therapy would be to use it in the same environment.'

'I agree. Valuable suggestion. I'd like to do what I can to help and will have the necessary papers drawn up. All you'll need is the agreement of the hospital and written consent of the patient, which is crucial.'

James nodded emphatically. 'Especially nowadays.'

John sighed. 'We must be careful even though the area is a grey one, the experts and government watchdogs are anxious to bring in tighter legislation for the control of drugs and medicines, not to mention the intervention of the legal profession. Pity we've to keep looking over our shoulders in case we put a foot wrong and when experts bombard us with paper work, most of which is gobbledygook. That last remark's between you and me.'

James replied despairingly. 'Understood.'

'All right. I'll supply you with serum and agents. Monitoring procedures will be outlined and don't expect any significant improvement for at least a week.'

John stood up. 'Come to the laboratory where most of the paper work's kept and the active agents. Please follow me.'

They walked along white corridors to the main laboratory – completely different -- to the Meetings' Room. It was cluttered with amazing equipment. James recognised several beta counters, light microscopes with ultra violet attachments, X-ray diffraction analysis machines and an array of computers with the latest backup equipment for manufacturing software. Also electrical equipment he'd never seen before. Bewilderment on his face was obvious as he glanced around.

John smiled. 'You may not be au fait with what we're trying to do.'

'Don't worry John. I've read a lot of your papers. They make good sense and I've heard you speak at regional section 63 postgraduate courses. The outline of your work is clear. You'll go along with the manufacture of chemotherapy agents but when it comes to testing you object to using them on sentient animals. These can be closely monitored on normal and pathological human tissue obtained, in the first instance, from patients.'

John's eyes brightened with surprise.

'Correct James in as far as it goes. Animal testing is too blunderbuss and complicated by many factors such as dissimilar metabolism, different physiologies, different genetics and so on. Now human tissue is available -- lots of it cancerous -- that we can bombard with chemical agents; the results recorded on computer and the reactions compared on normal tissue obtained from the same patient if possible -- if not elsewhere. Blood samples are also required at regular intervals to monitor the immune system. And I'm against using healthy human volunteers.'

James looked at him with amazement.

'Somebody should've thought of this before. Why have we to go a roundabout way when we've a simple normal /abnormal cell reaction, which can be measured precisely?'

John added. 'We're working on procedures that will modify conventional drugs in such a way to make target receptors on cancer cells attractive while making them less harmful on patient's normal cells. We're well on our way to solving the manipulation process -- not changing the agents only modifying them to be more efficacious.'

James hesitated searching for words – words he was dying to ask since he first arrived in Addenbrooke's but the chance had evaded him. Everything was technical and, somehow, got in the way of his inquiring mind about this man Nicholson. At last he took a risk.

'John. Please don't be offended if I ask what drives you along a lonely path and probably against a lot of vested interest that would like to put a stop to your work? Forgive my inquisitiveness – it may be none of my business. But I find it intriguing.'

John nodded and smiled pleasantly. 'Don't think twice about it. I've no difficulty in answering and the explanation goes back a long way. As I see it humanity is divided into those who do things and those who claim the credit. If you can do it try to achieve the former. There's usually far less opposition and rivalry.'

He turned and looked out the window over the flat Cambridgeshire countryside. Memories stirred sharply behind his dark blue eyes.

'It's too long to go into in detail. Since childhood I have always asked questions such as: why, where, what and when and always wanted to know how things work. My father was a teacher and I'm sure I wore out his patience many times. Still he did his best to answer. Some words of wisdom he tried to impart – something like this: "As you grow up and gradually pass from the vulnerable years of youth into the cruel and embittered world of manhood, make sure you retain your precious emotions and keep them intact. Don't discard them, or leave them aside, because you will find it difficult to harvest them later. As for forgiveness he who seeks it must be able to give it in return. A fool won't forgive or forget; the artless will forgive and forget, but the wise will forgive but not forget." Also let's just say courage. Just like that! To adopt courage as a way of life and have it as something to cling to always. I've learned that courage is not just a simple virtue but is at the centre point of every virtue at the crucial time and all aspirations and ambitions are permissible except those that climb on the credulities and miseries of others. You need it to build up strength of faith in yourself. In history it was the one thing I looked for in all my heroes – and when I found it I delved further trying to get an explanation. Of course, there are other fine qualities in life that are admirable – and there are all sorts of courage; the complete disregard for other peoples' feelings when you know they are wrong or merely trying to get you to do something unsavoury. Just to live as I have always wanted to, not because I have to — and know I've been privileged to be able to do so. James, some cynics would say it is better to be a has been than a never was – but that's not my philosophy. There is a lot more; maybe someday we'll go into motivation and adversity.'

He looked away for a moment and then added slowly.

'You know, James, I think the finest words in the English language are not "I love you," but "It's benign." There is a lot more.'

'I'll look forward to that. Thanks. It's been a real pleasure.'

He looked at his watch. 'James you've a long journey back to Welwyn. Let me get together our therapy pack so that you can start as soon as possible.'

'Yes the sooner the better.'

# 4

Karita Isselherg rarely permitted herself to be completely content – completely happy. Each day in Cambridge had been so frantic there was little time for such emotion. There was too much selection and elimination to go through in the ever increasing pace of daily life – full of clutter and confusion and location work was exhausting.

Things had changed since returning home to familiar surroundings; for once she yielded to this pleasant sensation letting it saturate every faculty of her being. No single factor contributed rather a multiplicity of items; her return to family roots, her recent engagement to John Nicholson and the near completion of The Trilogy. Her joy in loving John had brought a certain confusion – adaptation to life ahead would have to be worked out, the future opening up of vistas and horizons and offering, hopefully, a new and fulfilling life. Up to now she felt she didn't quite belong to it – unreal – just a dream. Now reality made her feel she did belong and gave a natural glow or sense of permanence.

She stood up from her desk and went over to the French windows. Everything surrounding her this bright morning seemed to add to her state of harmony -- the spacious dignified room with its discrete furnishings overlooking a large well-kept garden at the end of which a river flowed serenely under an antique bridge. A scent of seasonal flowers not only permeated the open window but enveloped her in an exquisite transient luxury.

She returned to her desk in the library of Kungsangen House outside Stockholm. Light filtered through the French windows and three candles flickered -- her style while writing. The flames, which made holes in the dark, fired her imagination and encouraged the flow of words. She'd worked consistently this way on The Trilogy and was nearing the end of the final part.

So far she'd covered ancient history and imagination was her greatest help. Even though the subject was mythology where fiction is linked to tradition she was acutely aware that good fiction should concern itself with differences in class. Observation was invaluable -- like a spark tossed into inflammable material; and writing on location was like hunting. There were times when she was buffeted by wind and rain with nothing in sight and she was despondent. Would she give up? The moment would open and there was something big, even stupendous. The process was mind blowing like in a spaceship with stars flashing past – amazing things had to be seen quickly and they were gone. This was a prize to keep forever. That's what writing was all about -- to her at least. Skilled writing was like a window pane.

In spite of differences in social classes it was her duty to deal with anathemas, to uncover, to speak the unspoken, and to reveal the interaction between what we say and what we do. Sensitive subjects could arouse powerful emotions and the ability to do so was a test of her skill with words. After all, most fiction was based on fact. The task was easier when dealing with ancient times and people, but as she edged towards modern mythology it could be a minefield. History got more opaque and dense as it approached recent times.

Several important sections were yet to be completed but in recent days she'd an urge to write the Epilogue. Unusual, but her mind performed in strange ways throughout the project and sections were not always in sequence; the impulse to complete the Epilogue was an example. She could learn a way to fill the gaps in the final composition. Something was missing, which would give it extra impact. She couldn't visualise what it was. As time was the best editor perhaps it would come with the answer and she'd recognise it when she saw it.

Anyway on to the Epilogue. She started to write.

# EPILOGUE

**...** Ancient Ireland was rich and remote, rich from trade and other dealings with Europe and remote because it was spared Roman invasion. Before the importation of Christianity it was peopled by Celts who were celebrated for their courage in battle, for their originality and beauty in art and for love of myths and legends. Their established customs were mainly oral and not written until many centuries later ...

... The point at which this book ends is where written history begins. Elsewhere one can read accounts of invasions, of clashes between chiefs and abbots and how an Irish Prince, with the blessing of a Pope, invited the English to conquer Ireland ...

... There are some who scorn imagination and regard dreams as a waste of time. Yet imagination and dreams form an intricate part of our lives because they allow us expand into countless other lives, of seeing beyond our physical senses dimension after dimension, through curtain after curtain to reveal a world of heroic deeds, passions and legendary achievements ...

... Sometimes there can be a thin curtain between reality and dreaming and time itself may bring the two closer together. Wells' book The First Men on the Moon was considered a tale of fantasy and yet we know men have reached the moon. And one has only to think of the horror of Hiroshima. Also in science and medicine the dreams of others are now being turned into reality at an ever-increasing rate ...

... The present work was a product of the imagination and told the story of a people who lived in a world uncluttered with the debris of technology, a people who were surrounded by a wonderland and where, beyond horizons of land and sea, the fantastical world of the imagination reigned supreme. The land of the Celts ...

Two thoughts came to Karita as she put her pen down and gazed into the candlelight. The sentiment about dreams becoming a reality focused on John's research. He probably dreamt his work would come to fruition some day and be recognised as the method of choice. He was working hard and it seemed to be developing into a full-time obsession. This worried her.

Another was her own work. The Trilogy wasn't finished yet she'd a deadline. The magic ingredient eluded her, which would make the end-product different. She'd have to be patient. Blowing the candles out she walked to the bridge over the river at the end of the garden.

# 5

Plenty of time for James Wilkinson to think on the way back to Welwyn, helped by traffic and hold ups. John Nicholson preoccupied him; he'd never met anyone so dedicated. What impressed most was John's demeanour and ability to listen with great understanding; he had one of those unusual smiles of reassurance rarely found anywhere -- full of compassion and empathy. Then as quickly as it came it disappeared and James was looking at a confident, tall, reserved colleague perhaps three years older than himself.

Even though the visit was short he gained insight into John's integrity and learned about attempts that could obstruct the research. John had no illusions and warned him to be on guard while treating Sylvia Clarke.

Next morning he entered the department and phoned Dr. Sinclair. Within minutes a knock on the door.

'Come in.'

Graham Sinclair breezed in and took a seat. He was a tall handsome man in his late twenties with a bronzed face and a shock of fair hair falling over a broad forehead. The dark penetrating eyes gave the impression that, most times, he took things seriously. Even though his height was impressive suggesting a successful rugby career, he stood erect giving him a commanding air of authority.

'I visited Cambridge yesterday.'

'Yes, that's my home city. My parents still live there.' Sinclair interrupted.

'Beautiful city. They're very lucky.' Wilkinson's voice was full of excitement. 'Nicholson's set up an intricate system for testing agents on normal and abnormal human tissue. You have to see it to believe it. I'd love to spend a week or two following him around.'

'Sounds great.' Sinclair was instantly awake. Any irritation vanished and he waited eagerly for the rest of it.

'Now to get down to brass tacks. We discussed Sylvia Clarke's case. It was decided she'd be happier here having treatment rather than a transfer to Addenbrooke's. Biopsies plus blood can be sent direct to them for analysis. Less upsetting and we've detailed instructions on procedure. It sounds simple and they are always at the end of a phone. With fresh material Nicholson can modify the agents to suit the individual patient -- it's almost a form of personalised medicine.'

'Very helpful,' Sinclair agreed.

'Impressive person. You'll probably meet him some day.'

Sinclair looked at him with surprise. 'Already have. I worked in Addenbrooke's. Remember?'

Wilkinson relaxed in his chair. 'Of course. Of course. My mind was on other things.'

'When may we start, sir? Sylvia Clarke is not well.' Sinclair asked.

'The sooner the better. We'll have to get the protocol cleared by the general manager and written consent from the patient.'

'And then we go ahead?'

Wilkinson did not smile. A disturbing thought was shaping itself in his head. He paused a little trying to find the words.

'Hopefully we'll get the all clear tomorrow. I'll work on the paper work this afternoon and have a gentle word with Sylvia Clarke.'

Sinclair gave a helpless shrug. ''Fraid she hasn't much choice now as things stand. She realises how ill she is in spite of our treatment and asked was there any hope? I didn't know what to say but did reassure her.'

'Well done.'

That afternoon Wilkinson had some difficulty with the manager in getting him to agree but after a personal phone call to his counterpart in Addenbrooke's he was convinced the project could go ahead. Wilkinson then asked Sinclair to be present when he spoke to Sylvia Clarke. She was sleeping when they entered the ward. A touch on the shoulder was all that was required.

She swung round and stared at him.

'Dr. Wilkinson. I wasn't expecting you to visit in the afternoon,' she said a little startled. 'And Dr. Sinclair too.'

'Please don't worry, Mrs. Clarke. We only want a little chat. Do you think you can make it to sister's office? We'd like a few words in private.'

'Of course doctor,' she agreed pleasantly. 'That should be fine. Let me put on my dressing gown and slippers.'

They reached sister's office and Sinclair closed the door gently.

Wilkinson made a gesture of resignation, picked up the hospital folder and after a moment shook his head. The X-ray viewing boxes glaring into the room were switched off. He sat down again, inviting the others to do the same but adjusting the light so that his face was in shadow and Sylvia Clarke's lit like marble.

She noticed this and understood. Each was an intelligent being and must maintain a detachment – especially the giver of news, usually bad, lest he expend too much of himself and eventually be left as frail as the patient.

She hesitated but felt strongly she should speak first.

'Dr. Wilkinson, there is something I should tell you before we proceed. I really didn't explain when I was rushed into the A and E Department two weeks ago. They're so busy and had no time for long histories, but what I have to say I feel is important and must be said to someone in authority.'

'Yes, of course, Mrs. Clarke. What is it you want to tell us,' he said easily.

Sylvia Clarke started by telling them she'd not been feeling well for months although she kept postponing visits to her GP. Eventually she was persuaded to seek medical advice. She was also having a lot of domestic trouble with her husband who would only listen to his inner voice about, someday, he was going to be a rich man when his gambling would make him an absolute fortune; the acute 'I. Me. My' syndrome left little else in his life.

She made an appointment with her GP who, after a short cursory examination, said.

'Nothing to worry about, Mrs. Clarke. Nothing at all.' He kept looking at his watch and mumbling something about how full his waiting room was outside.

'I think the main problem with you is " Four-Wallitis." It's not uncommon around here. You should get out and about, get a job and take your mind off yourself for a change. A woman of your age needs distractions, especially when your husband appears to have a one track mind looking at the racing results and playing cards. Now that's what I call unhealthy. As for yourself -- absolutely nothing wrong. Fresh air and getting out of the house will work wonders for you.'

She looked at him in disbelief.

'And just to help you get out of yourself here's a prescription to take the worry out of things. They have helped lots of my patients. You should try them.'

Her disbelief was increasing. 'I know how I feel and it's not right. But thanks for your help anyway. I mustn't take up any more of your valuable time doctor. Goodbye.'

She'd walked slowly out of the surgery. The late afternoon was lingering pleasantly around the shops in the main street. She idled in front of them. The cooling sun ended its promenade and called in its last beams from the narrow side streets. The pathways suddenly crowded with workers released from department stores and road traffic rose to a clamour. Buses were suddenly full with standing passengers – the unlikely few left on the pavement for the next full vehicle. To her this change was a sordid daily monotony as she quickened her pace to that apartment not far away she called home.

Once she arrived she looked out their sitting room window feeling lost – utterly lost and alone. The window faced south over the well trimmed outskirts of the city; the symmetry and tidiness of everything was a consolation when compared to the increasing chaos of her personal life; no children, an absent husband who spent some of his time in the bank because he had to and nearly always somewhere else, except home. She almost didn't exist.

Well, the doctor said she was all right -- absolutely nothing wrong. She drew a breath of relief for under her surface incredulity she half believed he could be right. A secret reassurance was beginning to unfold. Everything would be all right again in a few months, that her anxieties were just delusions and all she needed was to get away for a while to recover her inner equilibrium and balance. She'd savings in the Post Office that she'd hidden from Kevin. If he knew about them they would rapidly disappear down the bottomless pit of the bookie's large satchel and other reservoirs never to be seen again. She'd no doubts at all.

The doctor's words had woven magic qualities about her and with a timid avidity she temporarily yielded to a sense of returning life -- a life to be grabbed, to be lived. The rich temptations of travel -- not too far -- her budget wouldn't allow it but just enough to get away from the hell-hole of her current existence.

Certainly she would have to tell him – whenever she might see him during his infrequent visits home. The doctor said she was going to be all right. She couldn't go with Kevin – he had a job. Gradually and insidiously there dawned on her this could be the moment to let her see that things couldn't go on forever – nothing did – and with this new prospect of restored health a woman might have her own views, her own plans, might even think of marriage with someone stable, even have children, possibly somewhere near the sea. Her mind wandered into many possibilities of getting away from him, from here and from this place.

Then after a short while it all happened, fainting attacks even in the street and the ambulance was called.

Wilkinson did not smile, only looked at Sinclair for a long moment.

'That's quite a story, Mrs. Clarke,' he said softly, his eyes full of compassion.

She shook her head stubbornly. 'It's not a story at all, Dr. Wilkinson. It's perfectly true.'

'I'm sorry, Mrs. Clarke. I didn't mean it that way. I'm sure every word is true. However I should explain what we've been doing on your behalf in the last few days. On the suggestion of Dr. Sinclair here I've travelled to Cambridge and spoke to a Dr. Nicholson who heads a research team looking into producing safe and effective drugs for the treatment of cancer.'

'Oh goodness. Just for me. I don't deserve it. I'm getting treatment here.'

'You are indeed. But your condition is not improving as much as we'd hoped. You know that. Don't you?'

'Yes, things have not been too good. To be honest I feel worse now than two weeks ago.'

'We realise that, hence my talk with Dr. Nicholson. He's pioneering the production of new agents that are specifically targeted against cancer cells and yet have little effect on healthy ones.'

Her eyes brightened and she smiled hopefully. 'Sounds wonderful. But I thought your drugs were supposed to do that.'

'As you've just said you seem to be getting worse not better. Therefore the available drugs are giving you little benefit. Because of this we want to ask your permission to try you on the anti-cancer agents supplied by Dr. Nicholson.'

'Anything you say doctor. But I'm puzzled.' She prompted him sharply. 'Why do you have to ask my permission?'

'These agents are still in the experimental stage and are unlicensed for general release. These are the requirements.'

'Why is that?'

'The agents have not been tested on animals as recommended. And Dr. Nicholson disapproves of testing them in this way. Waste of vital time and also cruel.'

'I see,' Sylvia Clarke said and paused. 'I can appreciate that.

Animals have no choice. You are giving me a choice -- but what a choice.'

'You mustn't worry about their safety. They've been analysed rigorously using sophisticated screening techniques.'

'That's a bit complicated for me,' she said with a sigh, 'but Dr. Wilkinson if you recommend them I'll accept.'

'Good. I'm glad to hear it. Dr. Sinclair will explain the consent form. Then we can start treatment tomorrow.'

He stood up and put a hand on her shoulder.

'Thank you, doctor. I'm so grateful for your help and kindness.'

'We'll see how it works out. Now I'll leave you with Dr. Sinclair.'

Sylvia Clarke had little difficulty in agreeing to anything that held out a possibility of improving her condition.

The way was clear.

Next day arrangements were put in place. Monitoring equipment was brought to the bedside, double infusion lines set up and she was isolated in a room away from contamination. She was frightened and needed a lot of reassurance, especially in the initial stages.

Meanwhile Wilkinson and Sinclair sat in the consultant's office having coffee. Both looked serious and found it hard to initiate conversation -- unusual for them. Sylvia Clarke's account still bothered them.

It was Wilkinson who spoke first.

'You know Sinclair. Those GPs are not entirely to blame for missed diagnoses that are sadly too frequent. I blame the system they're forced to work laid down by bureaucrats who probably have nothing better to do than interfere in practice as it used to be run. Now the poor GPs are run off their feet collecting useless information. Once the data are collected it is probably ignored. Then those imbecile targets for GPs to achieve to be paid a decent salary -- no targets of increased patient numbers means no normal increase in salary and probably the reverse. But if targets are achieved there are generous bonuses for the managers. No wonder the system is rotten to the core. Miss diagnoses but don't commit the greater sin of falling below your target figure of patients and procedures -- otherwise penalties imposed.'

'Then you'll get some unscrupulous GPs milking the system achieving huge targets, pleasing the managers but leaving lots of human wreckages like Sylvia Clarke. My fear is that with these reforms creeping in by stealth managers will gain control of our in-patient practices and dictate waiting-times and directly affect the quality of service in hospital. I suppose the problem of managers is they will not or cannot trust doctors and nurses to do the jobs they were trained to do and given their lives for. It just makes my blood boil the unfairness we've allowed ourselves to become part of. Someone should put a stop to it. Cry halt! Allow us to do our jobs properly -- not looking over our shoulders at someone with watches and time and motion sheets.'

Sinclair remained silent throughout this tirade and then spoke.

'I think you've become set in your ways – to a certain extent. You're puzzled and afraid where all this will lead. We younger people with a conscience can vote with our feet. Just like administrators we believe the profession would collapse without us and some believe it until the day we die. It may not be true and other countries have shown the way. Even at this late stage you are fortunate you've been touched with dissatisfaction and not given up.'

He stood up and looked out the window as he concluded. 'I say you are fortunate because you have not reached that certain peace of mind, which some say, only comes when you have accepted the worst. However, no matter how long you live, the first twenty years are the most prolonged in one's life.'

James looked doubtful.

'No Graham, I'm not young enough to understand everything but I don't think that everyone would agree with you. Some would say that youth could be a wonderful time – if only it occurred later'.

Sinclair scratched his head and frowned.

'That's a little complicated for me. I'll have to think about it.'

# 6

Over the following days Sylvia Clarke had lots of time to think as she lay in the isolation room. She had assumed her marriage to Kevin would be a good one. After all, he was twelve-years older, a man of experience with a secure job in the bank. She was taken in by the sparkling blue eyes that looked straight at her and the abundant brown hair going grey at the temples. Maybe the good was in his circumstances and not the person.

Gradually certain failings came to light. He was brought up by a selfish mother, conceited carrying this defect into adult life. He was well aware of being handsome and somewhere he'd picked up a vague patina of good manners.

At first, things ran smoothly – on the surface. Gradually she realised her world was one of constantly making do and vulgarity she'd no experience. He'd said nothing to alert her about moving in with his mother who lived in an unclean apartment. It was merely taken for granted. She'd moved down the social scale. Although he'd a job in a local bank he knew no one. With these two people she was alone – an arrogant husband and a selfish mother, who disliked her, her demeanour, her articulate accent and resented the extra burden of living with them, which was supposed to be temporary.

Mrs. Clarke was a short fat woman, with pallid cheeks netted with multiple red lines. The hair, died brown, was short and straight and white at the roots. The eyes were small and never seemed to look directly at the person speaking; they wandered as if searching for something, and at her elbow lay a heap of old knitting and beside it a copy of 'The Tatler and Sketch.'

Many photographic enlargements of Mrs. Clarke glared in silence from the sitting room walls – shrill, arrogant and horrible. The real person continued to talk at Sylvia but with little eye contact as she was admiring the photographs or looking for invisible spots on the wall. Sylvia thought a sharp tongue must be the only implement that becomes more effective with constant use. The words of a sentence are usually inborn, she thought, but whining must be acquired. As the one-sided conversation developed Mrs. Clarke's laughs and gestures became more exaggerated and the room seemed smaller around them. She'd usually end by saying.

'Goodness, how time gallops, when one is doing all the talking.'

Eventually they moved into a larger apartment and the mother came with them to look after her son. Months dragged on in unimaginable dreariness made worse by raging tirades from Kevin gushing his own self-contempt on her, aided and abetted by the mother.

Nevertheless he was full of energy and climbed into some position of authority. He informed her he was a man of principle -- his reputation depended on it and yet he remained a bachelor at heart and jealously guarded his independence. They had no children, a fact that helped reinforce his bachelorhood. He liked being spoiled and the centre of attention, which meant a lot of duties and rules for her -- a form of slavish existence. He loved dealing with people, always convincing them he was right. He was one of those weak individuals, who generally attracted a patron and usually betray that person to a more important one. His marriage was also part of his work, which supplied the sentimental income and resented any attention she got from others.

She remembered one afternoon in particular and was taking

the doctor's medicine for a few days, which made her feel a little strange. Alone she sat in the sitting room as great clouds staggered across the sky outside. She settled in a heap near the window, hardly moving and appeared to be waiting for a visit or other happening. Slowly she drew one hand over the other, alternating several times, lips drawn tight and brows furrowed. The eyes moved slowly around the room seeing nothing.

She was confused. Still again with random thoughts; then she repeated the hand movements. She wanted to go away; to leave, to escape somewhere. This was repeated many times. She couldn't stand it any more. She was not living, only existing – never doing anything and always tired and listless.

Here nothing ever happened; got out of bed late, had a small breakfast and then almost nothing. She'd look for things to do but normally failed. She lingered during routine washing, mending and reading and then eventually to bed. This was done to fill the hours that dragged. One day led to the next with little change.

But today was different. No more mending, washing and cleaning. She was going away. Again she planned what items to take – the red dress and the warm grey cardigan. The suitcase was not large so there was little room.

With a certain purpose she climbed the stairs. No longer the drudgery of making beds and tidying rooms. With enthusiasm she packed the list in her head and at the tips of her fingers. She closed the case. It was not heavy. She counted her money several times. It was correct. Yes, she was going away.

Slowly she descended to the gloomy room – now for the last time. One final look around as she checked the mantelpiece clock. She'd plenty of time to catch the 5.10 train to King's Cross.

She quickly closed the front door and walked towards the station. Trees moved gently overhead and grass glistened with early evening dew. She moved with increasing excitement. She was now leaving all that behind. A new life ahead. Nothing else mattered. She was going away.

Her heart began to beat faster as her excitement grew. She hoped she'd packed everything she needed. Too late to go back now.

Anyway it wasn't important. She had her money safely hidden; this thought gave her security and she was overcome by a strange calmness and confidence. Everything was going to be all right. The oppression, the drudgery, the loneliness now was behind her – and finished. Gone. She was on her way to better things.

Her case grew heavier. She sat down for a rest and listened to the wind in the trees above. She wouldn't have to queue for her ticket as she'd the forethought to purchase it that morning. This gave her some minutes to relax in the knowledge of walking straight onto the platform. She repeated the rubbing action of one hand on another. The doctor's tablets made her feel tired but he insisted she try them anyway – so like a good patient she persisted. But her thinking was definitely slower and less clear.

The scene was surprisingly empty on the station platform. Perhaps everyone had arrived home from London – few were going in the opposite direction. Must be the case. Most commuters returned in the evening and she was not a commuter. She was going away. She walked up and down the platform with few about. How strange. There were no indicators for the next train and she began to worry. Still she was going away.

Impatiently she looked for someone. No one about. Then she looked at her watch – 5. 30. She felt a shiver – not of cold – of something else. The train was probably late. Still waiting, she found it increasingly hard to believe these words.

It was getting darker and the great clouds of the day grew larger and more menacing as she walked back. Wind also increased in sharpness as if mocking her and she'd to cover up tightly. The suitcase was so heavy. Before, these things didn't bother her – now she was full of dejection and failure, so tired to care any longer.

In the gloomy dark room she sat in the glow from the street. At dusk there was the slow drawing down of blinds and curtains all around her. The rest was silence. She tried to console herself – this wasn't the only day. There would be others. Someday. Yes someday. It all wasn't over.

It was too horrible. Too suffocating. She couldn't face it anymore – frozen anger was inside her. Again and again she rubbed her hands together subconsciously reassuring herself. 'Someday I will leave. Somehow. I just don't know how but I know I will leave here. It's too awful and unbearable.'

She bent her head and cried softly repeating: 'Someday. I'm going to leave and not come back.'

Eventually life was too much for her. She became ill. Suffocation at home bore down on her. She began to complain. This could not be tolerated and there'd be no two ways about it. Still things got worse. Kevin would not put up with insubordination and threatened to move out.

And here she was in hospital. Alone, fighting for life with the help of two lovely doctors and a stranger called Dr. Nicholson in Cambridge. Fate could be so unpredictable; maybe it was just as well. Sister entered the room to check her blood pressure

# 7

Word came through from Wilkinson that a committee of inquiry was being set up in the Q E 2 Hospital as part of another NHS reorganisation and Dr. Nicholson had been invited to attend. Initial inquiries told him that administration was looking into ways and means of improving the service. Then why Dr. Nicholson? He could advise on the running of his department in Addenbrooke's. No more information was forthcoming. Reluctantly he agreed to attend.

John was left to guess. When angry he opted for country walks around his home. It had stopped raining and he could probably make it around the woods before sunset. Karita was expected from Stockholm later and would make her way to the cottage. The path was heavy with mud, which made walking difficult. In summer it was one of his favourite walks but it was now in a sorry state -- a mere track separated by ragged bushes and occasional fences from the open grassland where wind and mist wandered freely without hindrance. He gazed over the scene where clouds only mattered and flocks of sparrows swayed in unison before dramatically swooping to the ground. Dusk rose from the earth and spread among trees, which looked like eerie statues against the sky. He moved on with the ground squelching and dragging at his feet.

Yet nothing like back to nature to calm the nerves. The sunset was red and similar to those in his childhood. The wide-open skies made for dramatic and glamorous effects and were an all-absorbing distraction. The fiery globe disappeared all too quickly behind the trees and hill and was replaced by the glistening evening star, brave, and assertive in the up surging dark, banishing the salmon coloured streaks of cirrus cloud. He sat on a flagstone overlooking a sloping field. It filled with mist along the ground; above was a rose line of red on the horizon and the North Star stationary above.

A rumble of traffic was heard, masses of purple oozed from woods, flickering of lights in cottages and smoke emanated vertically to the heavens. The city glimmered and shimmered in the south.

It was a beautiful evening, the parting day was calm and quiet as the wandering herds wound homewards leaving the world to darkness and to him. He loved the feeling of nature slowing, inevitably changing according to a predetermined plan. The way forward was destined and unavoidable.

This was his time of day. It afforded a peace and quietness and helped reflection. What did he really want? At this moment he was glad to get away from physical and thought traffic. Some friends had found the pressure too much, choked out their ambition and settled for an easier life. Yet the need to do the right thing tortured him, to achieve what he'd set out to do even though some regarded it as wishful thinking. Slowly, gently and silently the full moon moved through the night.

It was time to go. Stars had already established themselves as watchful keepers of the dark, the Plough sank into position and mist took up residence on low ground. It grew cold as he staggered along the path passing statuesque trees as if among friends. He slowly walked to the cottage. The harsh outlines of buildings, forest and hills were soft under the stars and silvered to an antique beauty. But he was blind to the blissful surroundings. The beauty he saw was in the faces and eyes of sick children and grateful patients. One thought flashed through his mind – one of the greatest problems today was not just cancer and AIDS, but rather a feeling of loneliness, stress and worry, and being deserted.

Karita was already home when he arrived. At first she was eager to greet him as she prepared food in the kitchen yet he was in an unresponsive mood. Then it came out; this damned inquiry in the Q E 2 Hospital to which he'd been summoned to appear. As if life wasn't hard enough -- now more intrusions and interference.

She busied herself. What was it that occupied her? Submissive poise with blonde head, hair hiding the face, made him draw into himself when he'd been so open to the enchantment of the night. Her lack of welcome troubled him. Perhaps she expected him to spoil and smother her and this did not happen. She probably blamed him. Disappointment welled up and he was reminded of the damp ground he'd been through. Strange how the Isselhergs were warm and welcoming during his stay in Sweden and how he got to know the family so well. Now Karita was different.

'What's the news?' She said eventually.

In angry tones he told her about the Q E 2 inquiry.

She tried to ignore his irritation and said. 'John. Go into the sitting room. It's more comfortable and I'll join you in a few minutes.'

Silently he nodded and did as suggested. He curled up in the large sofa and waited. She could see his profile from the kitchen; it was grim, yet vacant, and the body language indicated a huge degree of agitation. She placed a tray on the coffee table and started pouring tea and dividing food. She waited for him to comment and didn't look directly at him. He relented and smiled.

'It's really nothing to do with you, Karita. It's entirely my fault.' He sighed despairingly.

'That's not right,' she said still on edge. 'You're not to blame. We've been happiest when together. Long separations are not good for us and I'm to blame for that.'

He looked at her awkwardly. 'I don't understand Karita.'

She should tell him. It was only fair. She spoke softly. 'I've found the last part of The Trilogy the most difficult to write. It hasn't been easy. Even though I've finished the Epilogue there's still something missing with several gaps in the text.'

He mumbled softly. 'I'm glad to hear it's almost finished.'

'And I'm working to a deadline.' She appeared not to be listening as she continued. 'Then I'll say goodbye to the most painful phase in my life.'

He was amazed. 'Goodbye to The Trilogy?'

'Yes and good riddance.'

'You surprise me. I thought you loved it.' He said, suddenly serious.

'I did. But it was exhausting.' She searched for his hand. 'And in the meantime it has been keeping us apart. This makes me sad.'

'Goodbye is a strong word.' He frowned slightly. 'I hope I can be a suitable replacement.'

She laughed for the first time that evening, her expression full of yearning.

John withdrew into himself for a while, thinking; both had struggled in their own way against adversity, faced the same intellectual challenges, subjected to the same storms of doubts and disappointments. In spite of obstacles they'd matured, accepted failure when it came and tried again.

'I think you're beautiful,' he said quietly. The words were not as important as the caresses. She wanted them to linger.

As he held her close he whispered.

'Don't feel too badly about your work. It's going to be a great success. And I know life can be beautiful if you let it consume you in the right way. When you're working flat out and words are flowing as if they cannot be stopped, it can be glorious and fulfilling. Let it consume you like a fire, fed by a great draft. Once it slows down and begins to flicker then it can be depressing.'

'Life should be full of activity like that but other things have to be sacrificed,' she said with a hint of bitterness.

His face brightened immediately. 'No Karita. It needn't be. Sorrow and suffering can be part of a fire that fuels a successful life.'

She looked at him with gratitude. The words gave her courage and hope. She also guessed he was talking about his own battles and demons.

# 8

Tears and tension filled Sylvia Clarke at the start of the new treatment.

Later she was fully conscious as she sat up in bed proudly showing off a pink night-dress and lacy shawl as Wilkinson and Sinclair entered the room. A weak smile greeted them, her condition was so much in contrast to before treatment – it seemed almost miraculous. Her skin was still pale and a tremor was obvious as she sipped orange juice from a glass.

Things appeared different, staff more attentive or maybe it was the isolation, food tasted better, she rested more and time flowed quickly or maybe it was her imagination. She'd wondered about the outcome. She knew it was a risk – but what options were open? Very few. She had to go along with it and trust the medical team.

She smiled with greater effort as she put out a hand.

'Oh doctors, thank you so much. Thank you.'

The sister, normally serious and correct, was full of praise for the patient.

'It's the talk of the hospital, Dr. Wilkinson. Everyone's amazed.'

At the end of two weeks she looked forward to walks along the corridors; in four she strolled around the hospital.

Well, so far so good. Now, at last, after interminable months of getting nowhere, maybe, just maybe she'd the answer. Nothing was more precious than life, nothing more treasured than time, nothing more beautiful than the smell of cut grass, the fragrance of fresh blossoms and the joyous sound of bird song.

Life was good -- again.

Dr. Wilkinson was pleased with her progress; he spoke of letting her home. All she'd have to do was take maintenance therapy and was placed on the recall list.

*

Sylvia Clarke stood in the living room of her home with a conflict of emotions. Something was on her mind as she constantly rubbed her hands together. A frown came and went and came again. The room glowed with a flicker from the gas fire and electric light gave a soft yellow to everything. A letter lay open on the small table beside her. She picked it up and read it again. Her mouth tightened stubbornly. She went to the window, and opened one half to get a breath of cool night air. People passed in the street and traffic was light.

The stars were bright as she looked up and shimmering and seemed more alive, more real than the street. The letter dropped from her hand as the sound of a passing car startled her. She closed the window, and her heart was cold as the stars above.

Kevin wanted to see her again. His mother had died and things had changed.

The doorbell rang. Slowly she moved to answer it. Standing outside was Kevin Clarke.

'You have your own key. Why didn't you use it?' She asked sarcastically.

'Thought you might have changed the locks after I left.'

'What'd be the point? I haven't been well.'

'I know. I used to phone the hospital,' he said awkwardly, but meant it.

'But you didn't come to see me.'

'I thought you'd prefer me not to.'

'You're right. I didn't want to see you and still don't. So please leave me alone.'

He was insistent as he moved forward. 'No. Sylvia. May I come in? I won't stay long.'

She walked away.

'If you must.'

They entered the living room and he sat on the large sofa. She went to an easy chair beside the gas fire as there was a chill in the air.

'I see you got my letter,' he said quickly looking around the room. 'And you were waiting?'

What a silly question! She made no comment.

He forgot to ask about her health. This hurt but did not surprise her. He was thinner and paler and for a while was barely conscious of her, being preoccupied with his thoughts.

'Why have you come? Why bother?' She asked eventually. 'Have you suddenly become prosperous, found wealth again?'

'No,' he replied turning towards her, 'but aren't you even a little pleased to see me?'

'You never once visited me in hospital.' Her voice full of mockery. 'Of course I shouldn't have expected it after you walked out, leaving me to cope, alone and with nothing.'

'But you had your savings in the Post Office to keep going.'

'No thanks to you.'

'But it's true. Isn't it?'

She did not answer. There was no need. Instead she replied.

'Why have you come back and what do you want?'

She felt his desperation. This was ridiculous and she sensed unease as a spark of love seemed to ignite inside her. Or was it pity? She felt a little confused.

He sat on the sofa twisting a handkerchief in his hands. He was nervous and anxious, his whole attitude one of dejection.

'My life is finished Sylvia and mother's gone.' he said suddenly, almost to himself.

'I don't understand,' she said without interest.

He was angry inside. Really she didn't care about him, even if he was on his last legs.

'Don't understand what?' He asked abruptly.

'You haven't told me what.'

'I have liabilities, obligations and ...'

'Gambling debts you mean?' She snapped at him.

'Well, yes you could say that.'

'You could say that. So why don't you?' She was really angry. 'What's happened now? How much have you lost?'

'Five thousand pounds. That's all.'

'You foolish man. But your friends in the bank will help you out. So no need to worry.'

'It's not just the five thousand. It's my whole career in the bank. It's over now.'

'And your life is over for five thousand pounds,' she mocked.

'And my career.'

She grew more furious. 'Now it's only your career. I thought you said your life is over.'

'My life is my career.'

'Then as I see it you're not really a man you're merely a career.'

'I have my pride and my principles.'

'Have you now?' She answered with bitterness. 'I suppose that's what it means to be a gentleman, to have no courage outside one's precious career. Where's the man in you for God's sake?'

He shook his head in outright refusal. 'I have my honour, which is everything.'

'What has honour to do with everyday life of trying to cope, to make ends meet, trying one's best to live, no matter what may come? May I ask what your honour is right now?'

'You may ask,' he said avoiding her eyes, 'but if you're ignorant of that I cannot tell you.'

'Of course not. You couldn't.' A tense pause. 'So your debts have increased, not for the first time and now you're threatened with dismissal from the bank, therefore your honour is gone. Now what Kevin? Over to you.'

In complete command of the situation she spoke, not willing to give an inch. He winced like a boxer in a tight corner. He attempted nonchalance.

'OK. I know I've run up many debts and they may throw me out of the bank. After that there's little left. I could end it all. Or I could get a job as a waiter in a small time restaurant at a few pounds a week.'

'A lot of alternatives! Why not be a waiter?'

'Because it would not be me. That's why,' he said, evading her.

She looked at him, at his fine appearance, his obvious sensitiveness and she remembered his early years of proud upbringing and family background -- all too much of a come down.

'If they kick you out of the bank you'll land on your feet in something more suitable.'

This he could not accept. He'd a deep mistrust of himself and no belief in an isolated self-contained being. He was a clever, good-looking person and superior to other men. Apart from his family background he felt different. He envied the flexibility of the common workingman who could fit into any position with ease.

Sylvia could not understand these things. Perhaps it was best to leave her in the dark. The fine indomitable spirit, which a man must have in relation to a woman who loves him, was merely an ideal. He knew he was not it. He valued his attitude to this world; the world was his mistress more than any woman. He wished it were not so but sadly it was.

There was a long silence. What's the use in further discussion? A barren future faced him. She mocked and rejected him and her words had wounded him mortally. Yet a divorce was unthinkable -- the last ruin of his manhood. She would prefer him dead. The final solution to their problems because then her love would have nothing left to yearn for. There would be closure.

'If you decided to do away with yourself, go ahead,' she said in an emotionless voice.

'Probably the most honourable way out naturally,' he said weakly.

All was quiet again implying hopelessness for both.

'Why is it so terrible to come out of one job and look for something else?' She asked.

'I have my pride. Remember? And besides other men are not me.'

Why did she torture him? She appeared to enjoy it. He visualised himself in some low menial job and could not bear it. Who was she to dictate terms? She was the daughter of a commercial traveller, a tradesman. How could she speak for him, understand him? It was not possible. There was a limit to her understanding, a point beyond which she could not go. She'd better leave him alone. It was his fault, his crisis, his downfall and she should not get involved. But unfortunately she loved to mock him, to torture him and that hurt.

'Why can't you work elsewhere?'

'Work.' He almost shouted at her. 'What do you think I'm worth? A few pounds a week with some luck?'

He was pale, miserable and angry.

She tried to take control of the situation as a last attempt.

'It's not all true what you say,' she said, 'it's only your pride is hurt, silly fool you.'

'My pride. That's me all over. What am I without it?'

'You're a man, yourself. Again. Anything's possible.'

He shrugged dismissing this absurd suggestion.

'And our love we once had,' she added. 'It means nothing to you any more. I can see that by your behaviour.'

'Me, as a menial worker in a low paid job in the cogwheel of industry, what do you think I could bring to you in such a wretched lowly situation?'

'What does it matter?'

'Everything. It matters.'

'There's the kernel of the problem. It seems it doesn't matter what I feel or whether I have any say.' She was almost crying. 'They'll probably dismiss you from your important position in the lofty bank and you'll either be unemployed as a common little citizen, or you'll do away with yourself. It doesn't matter a jot that I am here.'

Indifferent yet listening he sat motionless on the sofa. She sounded vulgar. All this hysteria didn't change his attitude one bit.

'Can you not appreciate what value you put on me, you scheming little person,' she said, her anger rising. 'I have loved you, loved you with all my heart for three years and you deceived me when you said you loved me. And now what am I faced with? You'll do away with yourself because your puny vanity is wounded. What a fool you are Kevin.'

He looked with a fixed and superior gaze. 'All your words leave the bare facts unchanged.'

This incensed her all the more. 'Then go ahead for all I care.'

'There is one way out for me. You know what it is Sylvia.'

'After all this I haven't the faintest clue.'

He gave an ingratiating smile and said simply. 'All my immediate problems could be solved by a loan from you of five thousand pounds. My troubles would be over.'

'Oh dear. But for how long? I've heard that so often.'

'I promise it will be a make or break situation. If you could let me have the money all my problems will be solved and it would be like old times.'

There was silence. She would not give in. He stood up and, for some reason, said.

'God, it's hot in here. I'm going to turn off that fire.'

He bent down and clumsily twisted a switch.

'Is that your final comment Kevin? To turn off whatever exists between us.'

'Don't belittle me, Sylvia. I have my pride.'

She walked over to the door and opened it.

'You can keep your pride Kevin. Goodbye.'

He slowly picked up his coat and slowly walked across the room and banged the hall door with not another word.

She heard footsteps slowly disappear on the pavement. He was gone. Overcome in a suffocation of emotion she took her medication for the night. She had to lie down and felt she couldn't move ever again. Everything was at an end. She was finished, completely abandoned.

She lay there for a while as a sort of terror swept over her. Several times she attempted to get up but failed. Later she suddenly leapt to her feet in turmoil, a cold sweat on her hands. Previous hours, weeks and months were wiped away and she was taken back in years.

Why were they after Kevin Clarke, her old soul mate? He'd come to her because there was nowhere else to go. Frantically she moved around the room searching for her coat and hat. Yet to no avail.

An impulse beyond reason made her persist. With a sense of shock she realised why she should have helped Kevin Clarke. It was because it would be unbearable to live in an environment where there was no help – a world without care where anyone could be as alone as Kevin was this evening.

She collapsed breathless and exhausted on the sofa. Tomorrow was another day and would bring with it a welcome change to put things right.

Then sleep came over her amazingly quickly.

Outside the rising moon made soft shadows between the well laid-out houses of the city and Kevin's footsteps continued into the night.

# 9

The committee of inquiry had already commenced when John Nicholson arrived in Welwyn. Dr. Wilkinson insisted on being present with Nicholson and the committee agreed. The chairman, Mr. Edgar Butterworth, a fat middle-aged man with enormous owl-eyed glasses, was a striking figure of a man who might have been a good rugby forward in his younger days but now had the all obsessive hobby of golf – an obsession he could not get enough of. He sometimes said that the most difficult problems in golf were chicken-feed when compared with getting some doctors to agree on anything. He had switched from business to bureaucracy in his late thirties, and gave the appearance that he seemed to know more about medicine than most physicians – a lot of them were fools. Along with the striking glasses he had busy eyebrows that moved up and down at most remarks others addressed to him – his way of intimidation.

He welcomed Dr. Nicholson and then introduced him to members of the Board.

'Dr. Nicholson, we're most grateful to you for travelling all this way to help us with our deliberations.'

'Not at all.' He felt like a candidate at an oral examination. 'Although I'm puzzled as to how I can help.'

'Of course. There're many items on our agenda, which we won't bother you with. They're mainly concerned with looking at ways of improving the NHS.'

He nodded. 'One detail, Dr. Nicholson, which concerns us, is the rapidly changing field of chemotherapy. I understand you have a particular interest in this speciality.'

'Dr. Nicholson's one of our leading scientists in the field,' Dr. Wilkinson interrupted.

'Thank you, doctor. I think the committee is fully aware of that fact.' The chairman turned to John. 'We'd like to hear the current policy of Addenbrooke's in treating cancer patients. There are different approaches here that have left some confusion in overall policy. And I understand Dr. Wilkinson asked your help with a difficult patient who was not responding to conventional therapy.'

'That's correct chairman. My approach was discussed with him and I suggested a course of our in vitro treatment, which we use in some patients in Addenbrooke's.'

'The in vitro treatment was tried out in the Q E 2,' Dr. Wilkinson said, 'with the hospital's permission and the patient's consent and the results were successful so far.'

'How remarkable,' said the chairman, 'and to what do you attribute the success of your treatment, Dr. Nicholson?'

'Analysis of conventional agents is undertaken and we look for particulates that may damage culture lines of cancer cells. A study of the effects on healthy cell lines is also carried out. It takes a lot of computer technology to come up with the results of any drug. The most difficult part is to reduce the harmful effects on healthy tissue and yet maintain its destructive effect on cancer cells.'

'Can your work not be done on animals as seems to be the case in other laboratories?'

'No. They've a different approach to testing.' John was adamant. 'It's cumbersome, diverse, lacks organisation and has not got the minute and precise control we have. An animal's physiology is different to ours and has dissimilar ways of fighting foreign agents. We've greater control at every stage of the experiments and, of course, we only use human tissue in cell culture, normal and abnormal. What could be more direct? Although I would hesitate to use healthy human volunteers.'

Conversation started around the table and the chairman called for order.

'The entire field of chemotherapy is expanding at an enormous rate,' the chairman said, 'and there are strong views in favour of using animals for safety and efficacy reasons before allowing the medication to be used on humans. There is a powerful lobby.'

'I fully realise that, chairman, and animals have their place at a basic level of research,' John replied, 'some regard animal testing as going through all the proper channels. Nonetheless, for the reasons given I think that's old fashioned thinking.'

'I see.' The chairman saw a few worried faces. One member raised a hand. He nodded.

'Dr. Nicholson, I'm a member of the General Medical Council and we're at all times concerned with the efficacy of medicines and the safety of patients. Can you honestly tell us that your agents are as safe, or even safer, than conventional medicines?'

'Impossible to say at this time. But we are working on it by doing double blind controlled trials and we've excellent cooperation from staff and patients. It's just too early. Only time will tell.'

A lady raised her hand and the chairman gestured.

'Doctor, perhaps this is a personal question and you don't have to answer it. But is there any other reason why you are against using animals for drug testing?'

'I'm glad you asked that question. I was not going to bring it up. Let me say I'm against using healthy animals and subjecting them to procedures, which no doubt causes pain and suffering. The real question is not can they think, or can they communicate? It should be can they suffer? It is cruel. Where is the compassion? In vitro techniques are not cruel; they are clean, and most importantly, I will show they can be as effective, or more effective, than what we have available today. Someone once said that change is constant. I think those that will not accept new science must expect new perils, for time itself is the greatest innovator.'

The chairman smiled and moved some papers in front of him.

'Thank you, Dr. Nicholson. You have been most helpful and I might add quite convincing in your explanation. The field of chemotherapy is a grey one and is obviously going in different directions. I like the course you're taking in Cambridge. It has many attractions and, maybe, drawbacks. I'm sure this committee wishes you well in your efforts. Again thank you for coming. We appreciate your time and valuable comments.'

John nodded and smiled.

'It's been more pleasurable than I anticipated. I thought I might be in for a difficult time and have to face awkward questions. Not so. If there's any further help I can give please let me know.'

'We will, don't worry, Dr. Nicholson. We will. Good-day to you.'

When John Nicholson had left the room the chairman, Mr. Edgar Butterworth, turned to other members. His expression had completely changed.

'That person must be raving mad!' He scoffed. 'He doesn't seem to realise the sheer cost of producing a single tailor-made treatment for one patient -- the technology, the science, the utter know-how must cost millions. If that procedure was adopted for every NHS patient the expense would be astronomical.'

There was a pause. The GMC representative then spoke slowly.

'I would not be too quick to condemn him, Mr. Butterworth.' The absence of the word "chairman" was deliberate. 'If someone had said the same about the cost of producing the first phial of penicillin by Professor Fleming in St. Mary's in London and had the power to stop it on those grounds medical science would be in a sorry state today. Remember the saying about the wood for the trees? I think that Dr. Nicholson can see the wood very clearly that many of us just cannot see. I'm all for backing him in what he's doing and wish him well.'

There were confused faces around the table. Mr. Butterworth didn't appear to be listening as he continued.

'Nicholson is a bloody idealist and has his mind focused on one thing only. I don't know how my colleagues in Addenbrooke's have allowed him to get away with all this in vitro nonsense so far.'

'Mr. Butterworth,' said the GMC representative icily. 'Surely you are aware that Sir Kenneth Richardson is in overall charge there and has great influence on the University Grants' Committee. He would not allow a penny to be spent unwisely if he could help it.'

Mr.Butterworth looked a little subdued but by no means defeated.

'That's all very well doctor, but we have to live in the real world of now. And that "now" is getting harder and harder to even stand still. My God, I should know and I really don't think half of you gentlemen, and lady, have a clue how difficult it is—or perhaps you vaguely do and just wish to leave it to me and my management team. Those doctors seem to be continually looking forward to the past. I still feel the same and indeed a little frightened by what Dr. Nicholson has told us. Perhaps more people such as the pharmaceutical industry should know about it before it gets out of hand. Good-day to you all.'

# 10

The following day Mr. Edgar Butterworth had an important golf match arranged with three senior managers from his general team. The day started bright and clear and the omens were good for a pleasant eighteen holes. At the ninth a cloudburst occurred and from then on the match was a washout – all play had to be suspended and there was no likelihood it could be resumed at a later stage.

Frustrated feelings developed in Mr. Butterworth because these colleagues were special and he'd hoped to have urgent matters cleared up in the privacy of the golf course. Things were not going to turn out as he wished. Perhaps another time -- but it would have to be soon.

Instead he unexpectedly arrived home hoping to join other members of the family for luncheon. Not so. Emily, his wife, as usual was late. He asked the maidservant was she lunching elsewhere. The reply was Mrs. Butterworth had given no instructions for luncheon.

So Mr. Butterworth sat and waited for a hurriedly prepared meal of indeterminate fish and vegetables that did not appeal but, of course, he had a weak digestion mainly caused by living on his nerves. The food and entertainment budget for the household was considerably high because Emily had repeatedly told him she couldn't be expected to entertain her friends on dried fish and vegetables. Her social rating would plummet rapidly with fewer return invitations and that would be a disaster. She had pointed out that if he really wanted a quiet peaceful and dull life, talking politics to private friends he should have married a little harmless nobody. He should not have married Emily, an ambitious, confident and extremely popular hostess.

It was his fault entirely, of course. Although Emily did have some faults, which she recognised and secretly hung on to for good reasons. However, there were times when discussion of her merits, or faults, depending on whose side you were on, became tiresome. It did seem unfair, especially to Mr. Butterworth that life together should become one long adjustment to Emily's faults at the expense of his own. Oh well, he thought, moving shallow waters make more noise than deep flowing ones and tolerance was just a euphemism for indifference.

Emily made a habit of always been late which, she felt, gave her a sense of importance over those she kept waiting. She could develop frightful migraines if certain conditions became uncomfortable and also was adept at avoiding family visits and dull receptions; she easily argued with the servants who usually took it out on Mr. Butterworth. If ever a wife possessed the advantages and privileges these conferred it was Emily Butterworth. She knew exactly how to manoeuvre things to her advantage and gave her greater freedom from the drudgery of her husband's dull and boring job.

Yet in a shrewd and perverted way Mr. Butterworth had the good sense to see that he could benefit from his wife's wilful arrogance and that if these defects were the making of her, he could also learn from her – perhaps with beneficial results. He concluded it was unwise to be angry with Emily; after all he knew he was a reasonable man and would make efforts to vent his anger elsewhere.

For example there was Harold, their ten-year-old son, who was developing all the deviousness of his mother and barely tolerated his father. He was already finding fault with a greater range of household activities and Mr. Butterworth, in his warped wisdom, felt this was an achievement not to be discouraged.

Then Mr. Butterworth discovered that Harold had been receiving expensive presents from Mr. Richard Anderson. Now Richard Anderson had recently become a friend of the family and Mr. Butterworth prided himself on a high sense of propriety, had disciplined himself not without difficulty, to think that Anderson was a nice reliable chap and very charming. As long as he felt this way everything in his extensive house was perfectly in order.

And it was an expensive house – nothing but the best for Mr. Edgar Butterworth. Previously he had built up a fortune as a captain of industry, making a great success of managing people and production lines, which enabled him to build an impressive house in one of the best locations in Welwyn. As a result the governors of the Q E 2 Hospital thought Mr. Butterworth would make an excellent general manager and carry on the skills of best management practice in the hospital. A more than generous offer was made to him to run a large administrative department that was already top-heavy with experts but who needed skilled advice and guidance. Who better than Mr. Edgar Butterworth? So with a lofty golden handshake and generous shares in his company he set off to a new challenging life to make the Q E 2 even more efficient than it already was. And he was convinced that remuneration and bonuses of large businesses were merely a reward for success and achievement. It was a kind gesture by the individual to himself. And there was amazing zeal and dedication behind his move.

But the house he kept as his own. After all he earned it and deserved every luxury that was on open display although the various goddesses in various states of undress scattered about still caused a certain discomfort to his puritanical upbringing. Emily insisted they were seen to the best advantage in the great marble hall and he was not one to argue. The many paintings placed amongst them he did love and admired although again he was told that there were certain things he must possess in order not to be impressed by them; it was Emily he had to thank for that. She insisted that in well-to-do societies there was no discernable difference between extravagance and essentials.

Still she was not satisfied with the neo-Georgian house and extensive gardens. She'd reached that point where taking the exceptional as a matter of course and she decided she could not live without novelty and change. And Mr. Butterworth found this the hardest to accept about Emily; she rejected her successes as swiftly as her new gowns. Initially he'd accepted making allowances but as these allowances grew larger and larger his alarm also grew. One of Emily's major sources of strength was her obstinate obtrusiveness. Certain obligations simply didn't exist for her because she refused to recognise them.

The Butterworths had always been a haughty family and this trait had become worse after Emily joined them; it became more arrogant and conceited beyond logic. What didn't help the situation was that some years back they were presented at the Court of St. James's in London and ever since both remained in a state of permanent convalescence.

Then there was this charming Richard Anderson giving Harold, his only son, expensive presents. What was the point? Why did Emily allow it? Anyone who'd suffered Emily's condemnation of others not doing the right thing also would note that she could be inapproachable for long periods of time. So Mr. Butterworth decided to vent his wrath on the servants instead; then he remembered it was bad for his health so he took his medication instead.

Having taken some action he felt a little better, but not much. One of the spectacular marble nymphs smiled at him but didn't help because he knew how much these creatures cost and, worse still, was not sure they were worth the outlandish price Emily insisted on paying for them.

In his present state of unrest he avoided his study. It was always full of bills. In the library it was more relaxed not because of the books – more the peace and quiet. It was a mystery how some people would ponder and discuss and rarely argue over the merits and shortcomings of some volumes. He was glad to leave that kind of mental exercise to others as he was not well equipped to cope with lofty thoughts. Even though he took pride in having such a useless room it made him uncomfortable to sit in it for long. His own thoughts had little appetite for such matters or, perhaps, the apparatus to do so.

Gradually he drifted towards Emily's bedroom.

Hers, at least, was a room with a definitive purpose, an important part of the social fabric of the house and, of course, beyond. Mr. Butterworth knocked and with some trepidation entered. No one was there. He was delighted and even proud of Emily's popularity and the increasing number of social invitations she received almost on a daily basis gave him a good feeling. Everywhere he could see evidence of her exertion and toil. The desk was full of cards and requests as yet unanswered. Even the waste-paper basket was full. Looking down he noticed a rumpled letter at his feet. Because he was a man of fastidious habits and liked everything in its proper place he bent down and picked up the untorn letter from the floor. As he was about to place it in the waste bin he couldn't help reading the content of the note.

He sat on the bed and slowly continued to read on...

# 11

As Mr. Butterworth read he trembled on the edge of a storm, his thinking spiralling around in circles with this huge revelation. Looking around the four walls returned his stare in a cold, strange way. Who was this intruder, this stranger? What right had he to be here? Illogically he tried to hit back. I'm the one who bought and paid for you all with your silks, glass, chandeliers and the rest of the exotic rubbish. Why if it weren't for me you'd be as nothing on a shelf somewhere and now I could have you all torn down and returned to a God-forsaken rubbish heap where you belong.

He suddenly realised it was a large photograph of Emily he was speaking to directly.

He began to feel an unusual sense of power he'd not experienced for a long time. Of course, the room, the house and belongings were his. That was the complete measure of it. For so many years he was the person that Emily thought she could control and pour scorn on. He was the mere projection of her contempt.

Now he'd arrived. He was his own man at last. He could have told that imbecile, Anderson, where each one of her smiles could lead him. At present he derived a heightened joy out of his new found clarity of thought. The change to pure lucidity was an exhilaration. His brain was working overtime, planning, arguing and reasoning; all attempting to find the best and lasting solution.

Mr. Butterworth packed a few essentials and drove to one of the better hotels in Welwyn as soon as possible. His general demeanour had recovered sufficiently for him to appear outwardly normal. Still the storm raged inside leaving him with a feeling of almost pathological exhaustion. Nonetheless, normality began when such built-in phrases as 'doing the right thing,' 'the honour of the family,' and 'the integrity of the home,' came flooding back. He certainly was convinced he was right in leaving home immediately and said so in his note. He gave in to those calming moments of apathy when events hopefully seem to linger submissively at the door of the future.

The room he was shown into was in many ways abhorrent and foreign to his refined tastes. What would happen to his letters and business affairs? Where would they be sent? Would they be forwarded but to where? He had no idea. He must be going somewhere but didn't have a clue – as yet. He certainly wasn't going to stay in this neglected establishment any longer than he could help. London didn't appeal in this heat and the commuting; he was once more floating in a sea of confusion. His migraine returned and he slumped across the bed.

Time slipped by unnoticed and when his thoughts returned dusk was settling the mood for the night. Emily was probably in a state, or was she? She could be casually dressing for yet another dinner party with the pseudo-authentic excuse that her husband was 'called away' on urgent hospital business. You know how it is – these medical emergencies cannot run smoothly without the expert hands of the managers who always brought peace and quiet to turmoil and knew how to handle crises better than anyone else. After all, the whole system would collapse without them and their expertise.

Perhaps Anderson had been asked to take his place. The thought appalled him, but what could he do? He was borne down with the weight of indecision feeling like a traveller who had come to the end of his tether and needed to rest awhile – no matter where – even in this God-forsaken place. Still he was doing the right thing but for how long? Even his son, Harold, would have welcomed him at this late hour probably as a pretext to postponing bedtime. Mr. Butterworth felt domestically he had never been accepted more than a substitute for someone, anyone else. Not so at work he was in control and in constant demand. No decisions were too big, no problem was a problem for long and solutions came easily to him. This was good and life had a purpose in getting things done.

Now he was in a dilemma. He hadn't realised how much he missed home; Harold sobbing and confused, the servants in disarray with no guidance and yet he took some pleasure in the thought of Emily in tears and, perhaps, feigned collapse looking for sympathy but with few to give it.

A knock came to the door and before he could reclaim full control of his dignity the door opened and Tom Jeffers entered. Jeffers was a friend of the Butterworth family for many years. He used to be what is commonly described a 'wild man,' but now a reformed person who'd worked his way through medical school with enormous effort and cost to his parents and was now able to recognise suffering for what it was – mainly learnt from the destruction and mayhem he'd caused in his youth. It was also a source of satisfaction for him to be able to help distressed husbands and audacious wives. Mr. Newland, Secretary to the Board of Governors of the hospital, followed him. He was a serious man whose silences suggested a deep well of meditation and gave the impression he had something important to say even though he may have said nothing. He quietly acknowledged Mr. Butterworth's presence with the air of a prominent mourner at a funeral – one who intentionally overshadows the corpse.

Tom Jeffers broke the silence.

'Thank God we arrived in time.'

Mr. Butterworth looked around him with a mixture of puzzlement and pride knowing he was the centre of attention. There was a problem and through deliberate and concentrated minds that problem would be resolved one way or another – like a board meeting but much more important. He appreciated the presence of these two gentlemen but wondered what God had got to do with it.

Tom Jeffers continued his efforts at defrosting.

'We've come to have a little talk,' he said sitting down in an easy chair, 'about things...' and left them hanging in the air to be interpreted by each in his own way.

Again silence descended of its own accord until Mr. Butterworth, from a life-long habit of chairing meetings, inquired what exactly was on the agenda.

'Now Edgar.' It was Tom Jeffers again. 'I've seen Mrs. Butterworth. She asked to see me. You know?'

Mr. Newland, feeling he should offer some contribution, said slowly. After all common sense and good taste were his hallmark. Nothing more.

'Very prudent of her. I should think under the circumstances.'

Of course Emily had done the correct thing. Nothing less would be expected, but how much of this terrible tragedy did she reveal? Mr. Butterworth would have to thread carefully as he always did at board meetings.

'She showed me your letter,' continued Tom Jeffers. Not the letter on the floor! 'And she said she wished to apologize and cast herself at your mercy and consequently all would be forgiven.'

Mr. Newland tried to demonstrate his conciliatory skills as he interjected a note of absolution. 'That should be held in her favour, Edgar. It's an awful damned business. Poor Emily is devastated and broken. It's evident to all, although they don't know the reason why.'

'As friends of both of you,' said Tom Jeffers, 'we've come to see what you intend doing.' Mr. Butterworth was upset again – the burden of decision was again placed heavily on his shoulders. And yet he was strongly convinced that there was no one more worthless when riddled with indecision.

'I intend to leave her. Go away. Forget about it.'

'Divorce her you mean?'

'Why yes. Yes.'

Jeffers looked surprised.

'You're sure this is what you really want?'

'Yes. I do intend to do exactly that.'

'You've decided then?'

'Yes.'

The repetition seemed unnecessary and there was almost no need for further discussion, but like any board meeting things had to be looked at, considered, discussed, options covered and other planning arrangements that needed to be resolved. He was about to go into chairman mode again when Mr. Newland interjected.

'This kind of thing can be bad publicity in the business world. It can create all sorts of prejudices.'

Mr. Butterworth almost scoffed at this remark. What sort of good had his money ever done him? All his hard work, long hours, worries and anxieties only served to make Emily a rich woman. He had little to show for it and certainly not enjoy it.

One last attempt came from Mr. Newland.

'And what about your son – and any further children? Surely they must be considered, planned for, taken care of, or what you will?'

'Harold has never considered me.' Mr. Butterworth retaliated vigorously, implying that was the end of the matter. He now felt he had reached the last item on the agenda; any other business. His spirits rose. It was now his turn to exhaust his listeners with all and every wrong he had been made to suffer for such a long time – not just this final catastrophe. He felt that Emily would, at last, learn about all the injustices and failings and other injuries he had borne with silent fortitude.

He wasn't finished when Mr. Newland glanced at his watch and looked at Tom Jeffers. They stood up reaching for their coats with an air of resignation. They had come to the limit of their duty, even their endurance.

The meeting had been concluded.

# 12

Kevin Clarke woke in his lodgings with a feeling of despair. He remembered what a fool he'd been. She'd even called him a fool! He couldn't escape his debtors, which were quickly closing in like vultures. The misery of his situation was alarming. Nothing in the future and his last hope, Sylvia, refused to help.

He tossed on the bed; the small room seemed to smother him. A gust of wind rushed through the bedroom blowing the curtains in at one end and out the other and twisting them like a rope up towards the cracked ceiling and brushed across the wooden floor – and then was gone. The air felt heavy drifting over him carrying a shaft of sun laden with all the cares of the world he'd have to face that day. His debts appalled him and when they came to light he would be dismissed. He'd lose everything. What then? Only one solution -- a hideous one. To do away with himself seemed the only answer. There was no other solution. And she could easily have saved him. Let it be so.

At work an overcast sky developed necessitating lights to be switched on prematurely. It was oppressive. After work he decided against public transport and walked past the old home on the way to his lodgings. He reached the suburbs with its maple leaf trees along the wide streets. The sulphur lights gave an unearthly glow to people, shops; even the children were pale as death.

As he neared the house there was unusual activity. People were gathered around the entrance. Oh God no. Not more trouble please. No more. He could not face it. Ambulance men pushed a laden stretcher into the van, shut the door and moved away.

He went up the steps into the hall.

A police officer was making notes in a small book. He turned and faced Kevin.

'Good evening, sir. A sad business I'm afraid.'

Kevin looked shocked. 'What's happened?'

'A young lady found dead in the front room,' the officer said gravely.

'Dead,' Kevin repeated eyes wide with fright.

'Yes.' He looked at his notebook again. 'Apparently suffocated by fumes from the gas fire. Although it'll have to be confirmed by the pathologists. What a terrible accident. If people would only be more careful this kind of thing wouldn't happen.' He gave a gentle cough. 'Did you know the deceased, sir?'

'She was my wife,' he continued, indicating with his eyes the room and surroundings, 'but we separated three months ago and have lived apart ever since.'

'I'm sorry to hear that, sir. But if you don't mind we'd like a statement from you whenever convenient. Could you possibly call to the station sometime tomorrow so we can include it in our records? Poor girl. What a way to go. I believe she'd been making good progress in the hospital for a serious illness and it was well controlled. By all accounts she'd everything to live for.'

'She left no note or anything suspicious?' Kevin asked a little breathlessly.

'No nothing like that. The forensic people have already gone over the place with a toothcomb and found nothing unusual. For the record I presume you're the next of kin and legal matters will have to take their course.'

'I'll give you my name and address and, of course, I'll cooperate in any way I can.'

As Kevin Clarke left the house, walked down the steps to the pavement and on the way back to his lodgings a smile appeared on his face and remained there. No such expression had occurred for many a long week.

# 13

It was warmer than usual on the day of Sylvia Clarke's funeral. The small chapel was about quarter full of friends and neighbours and her husband Kevin felt he had to attend; John Nicholson wanted to be present although the two men had never met until that morning. Flashes of Linda's funeral came back giving him a mixture of sadness and relief; sadness at her loss and relief that she suffered no more. She spoke to him in vague terms again and it was good. She whispered that it was wrong to go back – not even attempt it. The true meaning of life is going forward. It is a one way system and the right way is travelling in the direction of your fear.

It was a short ceremony and the mourners moved slowly to the local graveyard. The black hearse passed through the gated entrance and along a dusty path edged with quaintly carved cherubs lying in sodden sleep on stone pillars. Occasionally a kneeling relative with fresh flowers was seen praying, but silence lay over most of the graves with leaves giving their scent of shadowy memories to passers by.

As John walked back to the hospital the heat grew giving a drowsy picturesqueness to shrubs and trees -- never unpleasant, only comforting like a warm blanket for the infant earth.

*

New ideas flooded into the experimental laboratories in Cambridge. Techniques for obtaining normal and abnormal human tissues were developed and carefully processed. A patient could be admitted with a primary liver cancer; other centres would have routinely administered a cocktail of agents tried out on animals.

Nicholson's approach was different. He acquired healthy and cancer cells from the patient, grew each separately, subjected each to various agents and got immediate results of cell damage. This was then measured and the agents specifically modified.

At the time, this was new thinking and by-passed months of laboratory testing on healthy animals. His method was quick, direct and logical and the ability of his computer teams to detect minute damage at cellular and subcellular level was ever increasing. A study of the intricacies of cells about to be treated must be the way of the future. The system was direct and obvious -- to him. There should be no other way.

# 14

Tom Nolan continued his painting journey from Malin to Mizen. After all the glories of Connemara, Coral Strand, Ballyconneely and Costelloe he travelled through Galway City to the Burren in Clare. His reaction was bewilderment; stones and pavements were everywhere and even stone mountains in all directions. Walking through stone fields he felt like a foreigner.

Sky had always fascinated him and here it was open and vast – clouds could sail in close formation with the curved hill crests and birds buffeted to places they didn't want to go. In areas people had exercised an amazing sense of belonging when a few stone acres were surrounded by walls, breast high, and the land covered with a mixture of seaweed and soggy grass, where tough looking sheep, perpetually optimistic, sought out nourishment.

Overlooking this was the vast loneliness of the sea, but a loneliness that wasn't unattractive, and its undulating surface and hypnotic movement encouraged one to look for signs of life – human or otherwise.

Tom searched for suitable locations – but they nearly always evaded him. As he trudged along through pools of turf brown water gathered in his heel marks. Each footfall expressed from the bog the rich smell of autumn.

Then an amazing sight appeared -- a turlough in Clare with the midday sun reflected on it and he had to paint it. He'd discovered a rare lake on this moonscape and he'd the good fortune of capturing it on canvas before it disappeared again below a grass field. Colour was everywhere, in sea and clouds and especially in wild flowers.

At the end of an exciting tunnel on the Chaha Pass he saw an unbelievable never-never land. The mountains of south Beara including Glenlough, Sugarloaf and Hungry Hill were veiled in a pale unearthly light. Near the top of Glenlough Mountain was Barley Lake shimmering brilliantly through the mist. It dazzled and moved about and the more he looked the more it seemed to overflow with light -- beyond the paint brush and also, perhaps, beyond words. A land of fiction whose experience burned itself into memory like a flaming arrow.

He arrived at Mizen, the southernmost point of Ireland -- Land's End. The seven- hundred foot rock face was marvellous and menacing. The final painting was of the rocks and cliffs surrounded by screaming gulls and thunderous waves pounding rocks far below the tiny footbridge.

When Tom Nolan came back from his location work he brought with him an air of rejuvenation not entirely due to changes of climate but to a certain companionship that accompanied him on his travels. To Ruth's ever-sharp eyes he had started out alone; but in reality an irresistible and welcome companion had also travelled with him; this agreeable companion just happened to be an all embracing idea.

Anyone who has ever tried the encounter may have experienced its exhilaration. The intellect, once developed to the dimensions of greater ideas, never returns to its previous size. During these times Tom would not have changed places with the greatest victor in battles and conquests and even the most ravishing female tended to be accompanied by baggage along with strict and rigid inhibitions that could be a major encumbrance. Whereas an all embracing idealism did occupy his nervous system to such an extent that nothing else mattered. It could start in the smallest part of his brain and expand to encompass the whole horizon. And it was completely flexible occupying him on a train journey, car travel or on foot over a treacherous bog and they would communicate in joyous complicity.

From the outset he found his companion and friend – idealism – the most agreeable of fellow travellers. However, it was really in scented and redolent solitude of woods and the glories of sunsets that he enjoyed the full meaning of his intimate companion, drawing the entire measure and sparkle from deepest thoughts.

And the resultant work on canvas would reflect this overflowing ebullience with reality.

His mission completed he'd to return to his family and studio in the cottage beside the harbour in Greystones. A lot of work still had to be done to prepare for the exhibition of this part of the journey. Elation and exhaustion filled him as he drove back home.

Weeks prior to the exhibition were hectic. Paintings were completed, signed and varnished yet something was missing. Another brochure should be written with illustrations and maximum publicity obtained. And who should open the exhibition? Normally, a well-known celebrity was invited to open a private view and give a short speech to guests. The gallery and family were stymied on this; all along Annette knew who would be her preference. It had to be Dr. John Nicholson from Cambridge, the man who'd saved her life. She suggested it to her parents and they thought it was a brilliant idea. That is if he would accept.

That night Tom Nolan drafted a letter to Dr. John Nicholson.

# 15

One morning a large envelope was delivered to Dr. Nicholson's cottage. He found it on his return from work. The address was hand written and he recognised the writer's flowing style. Yes, he was right. At the back was printed: From Tom Nolan with details.

He opened it expecting unwelcome news about Annette; but no, it was about Tom's painting activities and the journey from Galway to Mizen. What a coincidence. He and Karita had been to Mizen, which had left an indelible impression, especially on her.

Towards the end Tom diplomatically informed him that he and the gallery would be thrilled if Dr. Nicholson would be the guest of honour and open the exhibition by giving a short talk in any manner or form he so desired.

His first reaction was to say no; he could not do it. He wasn't an expert on art and could imagine press people and critics hanging on to his words. Out of his depth. Surely there was someone more suitable.

Karita entered the room and noticed his expression.

'What's the matter John? You look worried. Bad news?' She said, suddenly serious.

'Not bad news at all. But still a dilemma.'

'How do you mean? That sounds strange.' She asked, unsuspecting.

'Do you remember Annette Nolan in Addenbrooke's some months ago when I took part in a marrow transfer, which seemed to go well.'

'Yes. And I advised against it because I was afraid things could go wrong for you.'

'That's not the problem now. The father, Tom Nolan, is quite a famous artist and is preparing his second show of a journey from Malin to Mizen. He's even produced an impressive brochure about the journey and enclosed it with this letter. You might like to go through it. He ends at Mizen and has some remarkable paintings including the view from the suspension bridge.'

'Don't remind me!' She replied, her eyes full of memories. 'I was scared out of my wits and felt something strange going on in that place. I must study it. There may be some ideas in it that could help. There are many gaps to fill. Getting a different point of view could be helpful.'

'Take it and study it. It'll mean more to you than me and,' he smiled, 'who knows, if I accept, you might help with the speech.'

She was surprised. 'Am I invited?'

'Of course. Both of us. We can look forward to a few enjoyable days in Dublin and I'll show you some places worth seeing.'

'And keep me away from those not worthwhile.'

'Quite right. There are plenty of those.'

John read the letter again and felt a thrill at the prospect of addressing a group who would probably be outside the medical profession but important in other fields. A new experience.

Before accepting he'd set aside time to think about the meaning of art, the pros and cons, yet not get into difficult water like the trouble he got into with Linda in Stockholm's Museum of Modern Art. He objected to an old boot tied to a canvas on a wall, which had received praise from the critics. It was complete rubbish. He thought everyone is capable of expressing far more than we actually allow ourselves to do. We all know more than appears on the surface. This is probably because of a certain fear in the face of expressing our inner thoughts in an incoherent manner that could result in ridicule or even hostility in others. We should not have this fear when we wish to express our true feelings about certain forms of trash that purport to be art masterpieces by a certain group of people that have ulterior motives behind what they say and write. We should stand up and be counted.

Anyway that was another time. And Tom Nolan's work was pleasant and exciting -- the complete representational artist -- for whom he had admiration and sympathy although he'd have to find out more about the meaning of art.

# 16

At breakfast next morning Karita was full of enthusiasm for Tom's art. John was already halfway through his meal and preparing for work. He'd time to listen.

'You know John, even though the writing is somewhat stilted his personality and enthusiasm shine through. He's a love for places he visits and a sense of history.'

'Could be useful when selecting a scene for more than the view. Perhaps he's able to see between and beyond the surface of things and bring out the true meaning of landscape.'

'Exactly. He has empathy with people who may have lived in a place perhaps hundreds of years ago and, sometimes, it has changed little, which gives him great delight. One sentence he uses I think is brilliant. It goes: "This has been a journey through spaces and places, searching for those mindscapes that no one has seen, in landscapes everyone has seen, and resulting in memories intended to implant an awareness of the deep interdependence of the world without and the world within for, surely, it is through art that our world is made visible." '

'I like that. It summarises a lot of things and I may use it to finish off the speech.'

She nodded. 'Have you started it yet?'

'I scribbled some notes last night when you went to bed -- rough notes mind you.'

'The offer is still open if you'd like me to do a bit of fine-tuning -- like for the Oxford Union. I'll be more than glad. Will you remember that?'

'I will and thank you. We've another couple of weeks to go. Depends on how much time I can devote to it in between my work.'

'Tom Nolan sounds like a fascinating character, intelligent in his own way with fixed ideas and strong ideals. I'd like to meet him some day. There are questions I'd like to ask him about my work; to a certain extent it runs parallel and I could possibly learn something. If that's all right with you, John?'

'I can see no reason why not. You've to explore all avenues to produce the best work for The Trilogy and Tom Nolan is certainly an original, if not tempestuous, source. He may be difficult to handle at times. I had several rows with him a few years ago on entirely a different matter. It shouldn't affect your work in the least.'

'Thank goodness for that. You had me worried for a moment.'

'No need to worry. At the time he was campaigning against using animals for experiments and one thing, I think, he said was: It takes forty animals to make a fur coat and only one to wear it! But painting is now his main occupation. The family are wonderful and they have a lovely daughter, Annette, that I got to know in Addenbrooke's months ago. She's a remarkable child who'll develop into a real beauty. Knowing what a perfectionist Tom Nolan is I'm sure he'll have everyone running round in circles. It should be a good show. He deserves it.'

'Looking forward to meeting him, John.'

# 17

The day of the exhibition arrived and the Nolans met John and Karita at the airport with hugs and kisses. It was mid-afternoon and they drove to the city centre hotel. Tom would call for them about 6.30 p.m. The private view was scheduled to start between 7.30 and 8 o'clock.

Tom invited John for a coffee in the lounge before going to the gallery.

'Forgive me Dr. Nicholson but … '

'The name is John.'

'Fine then, John. Forgive me for asking but are there any questions you want to ask about the show or my views on painting in general?'

'You had strong statements in your brochure about the last chance saloon for artists, changing landscapes for the worse by the greed of the JCB, interpretive centres located in unsuitable places, careless farming ruining the environment -- even the voracity of gold prospecting in sacred places and the road and rail services destroying the very things people come to see. I can't touch on these as you've already done so eloquently and other things besides, many of them emotional and beautifully put. You've left little for an outside commentator to cover.'

'John, I completely disagree. You're presence tonight will add enormous prestige to a society that thrives on glamour. People will be watching one another, and their reaction to your words. This is a city of interval watchers where it's almost more important to be seen than to come and see. They're a breed of their own.'

'I've a good idea of your views from the brochure.'

'Oh that,' Tom shrugged. 'That's just formal guff. I had to say something about my world and to keep it fairly low key about a few people's efforts. Some of it's diabolical. Think of the shameful nonsense of the Turner Prize. Pitiful rubbish! As Brian Sewel, the famous London art critic has said: "I don't know what art is, but I do know what it isn't." '

'Your work is so detailed and finished to such a high degree. Does it not slow you down on the production process?'

'Yes, John. You've put your finger on it precisely. Thank God, I'm not part of the bucket and slosh brigade that sell their junk to the Arts Council and then is splashed all over government buildings. There are good artists out there who could do excellent work, be commissioned by the OPW, that would be more acceptable to the general public -- the real people -- and not a bunch of experts who are employed by the Council because they can talk in convoluted language no one understands and probably not even themselves. It's such a sorry state where private artists struggle on their own against a flow of experts who really know nothing, advising civil servants that they should keep up with international standards of ugliness and depravity.'

John sat back in his chair and smiled.

'Strong statements Tom! Tonight a little diplomacy is called for. What I'm going to say will be my personal reactions to your beautiful paintings and try to get people to see my point of view. It will complement your work by a relative layman who can make an educated guess at what the general public would like or not like. We'll see how it goes.'

Tom looked at his watch. The gallery was within walking distance of the hotel.

When they arrived Tom asked. 'John would between 7.30 and

8 o'clock be all right for Adrian Osborne, the owner, to call people's attention for your address?'

'That would be fine.'

'In the meantime Karita and you might like to take a stroll around the gallery and look at the pictures in relative peace before it gets too crowded.'

'Excellent idea,' John said. 'Come Karita. Let's start at the beginning. I see each picture has a few lines underneath to explain details.'

For the next twenty to thirty minutes they studied all seventy-four exhibits mainly oil paintings and a small number of drawings. Towards the end it became difficult to get a good view of some of the paintings because of the build up of people and the flurry of gallery staff putting red stickers on pictures even before the opening speech. Karita was occupied making notes about some of the works.

Eight o'clock came. Tom asked John would it be all right to proceed.

John smiled and nodded.

Adrian Osborne clapped his hands. Silence broke out. He cleared his throat as he approached the microphone.

'Distinguished guests, ladies and gentlemen. It gives me great pleasure to introduce one of our finest experts in the field of medicine to speak to us tonight. He's travelled from Cambridge University and kindly accepted our invitation to officially open this exhibition of paintings by our very own Tom Nolan. Tom's second exhibition of paintings is a result of an epic journey from Malin to Mizen. They represent a volatile landscape not only in geography but also in our rapidly changing way of life. And these changes are not always for the better. What Tom has done is to look for the most precious sights and scenes that may not last forever – perhaps only for a short time -- so that we can have a permanent record of things disappearing. I've said enough. Let me introduce Dr. John Nicholson who is Head of the Oncology Department in the University of Cambridge and in Addenbrooke's Hospital in England, yet is an Irishman who has made an international reputation in the field of cancer research. We can all be truly proud of him. Please welcome Dr. John Nicholson.'

John stepped forward, reached into his breast pocket and took out some reference cards to help keep his thoughts in order. Raising the microphone to his height he paused as he looked around at the assembled guests.

# 18

'Good evening ladies and gentlemen. It's a great pleasure to be here this evening. I've been asked to say a few words to officially open this exhibition by Tom Nolan. Well, I've come here tonight not just to say something, but hopefully because I've something to say. Yet I wonder why me? I'm no expert on art although I know what I like and what I don't. Perhaps it's the same for most people. If so we share something in common. Looking around the gallery earlier I certainly liked what I saw. It was fascinating and wonderful.'

'Tom Nolan's representation of the west coast of Ireland is more strictly speaking a re-presentation and a journey of discovery. I believe his central theme is to try and convince us that what we see with our eyes is not necessarily what we create as art.'

'Tom's journey must have started some time ago when he developed his own ideas about art. Many years ago it was drawing that started things off; the mastery of the discipline was essential to establish basic principles. There was a parallel development in painting. Understanding both allowed a correct balance to be achieved between the method and the message in his painting. It's obvious artists must have something to say, to communicate. Method will affect the message in that if the method is not adequate then the message fails and the work fails. Tom is well aware of these facts and has caught the balance beautifully.'

John looked at the two paintings behind him and then at the audience.

'As a medical doctor I often think this relationship is similar to the workings of the human body. The skeleton and flesh can be likened to method and message. To explain further let's consider the anatomy of a painting. Bone structure and composition are essential foundations before we breathe life into it. Then the physiology of a painting implies the correct laying on of structures in the right place and in the correct proportion so they can work in harmony. Pathology, unfortunately, has to be considered sometimes.'

He looked at more paintings to his left and right.

'But I can see no pathology here tonight.'

Someone clapped and said. 'Here. Here.'

'Yes. There are some sick pictures around today but not here. Finally, the psychology of a painting is, of course, the emotional content. Does it move you, or leave you cold or even make you angry? If you know the answers to these questions then you are a long way to understanding art. Perhaps success in painting is using structure and colour to produce in the viewer the same emotion that moved the artist.'

He paused for a moment and glanced at Tom and Ruth.

'Staying with psychological values -- values common to all of us, our sympathies, our concerns and those active at a subconscious level -- dare I use the words philosophical principles? In other words the artist's ability as seen through technique, visual opportunity and insight can be evaluated as one who sees and feels, not only what is before him, but sees the scene in its universal implications discovers and unfolds the one in the many and the many in the one. I think Tom Nolan has a special gift in that there is a pathway from the eyes straight to the heart and bypasses the intellect. Ideas can best be conveyed through words and the artist can then communicate his emotional reactions to them in paint. I think Tom Nolan has done this superbly throughout his journey.'

Spontaneous applause occurred at this point and John had to wait until in died down. He raised a hand.

'Ladies and gentlemen, I'd like to finish with a few personal points. In my view landscape painting has always been a great deal more than the mere depiction of Nature. The impressionists were fascinated with the effects of strong sunlight. Then an increasing preoccupation with materials eventually led some to exclude all or most recognisable matter and we had modernism, which at times was carried too far by the terminally self-satisfied. But in the last ten to twenty years the rebirth of realistic painting thankfully has occurred. Unlike previous periods there is now a multiplicity of ideas, attitudes, styles and directions.'

He went on to criticise the photograph as an inert, a dead presentation carrying no emotional baggage and no real meaning.

'We are all reassessing Nature for very good reasons. Tom Nolan has a personal vision that underlies some of the ideas of contemporary realism and contains a rationale that all our latent and universal fears of a world threatened with nuclear contamination and global obliteration, and an environment threatened with crime and over population, can find refreshment light and peace in a reappraisal of Nature. Knowledge of these things is essential for an appreciation of all that is good and noble around us. With it we will not only understand others but we can better understand ourselves. Tom would certainly agree with me that the central theme tonight is that creative art is not necessarily the same as what we see with our eyes. It is more. It has to be if it is going to work. What the "more" really is I cannot answer in words. Seamus Heaney talks about "poetry taking the strain." Perhaps by way of answer we can allow what are present on the walls around us speak for themselves and, who knows, they might even take the strain.'

Enthusiastic and generous applause greeted John as he finished speaking and a clamour of conversation broke out. Tom and Ruth came forward to him obviously delighted at what he'd said. Tears were in Ruth's eyes and Annette tried to hold his hand; he gladly accepted.

# 19

Tom Nolan invited John and Karita to a meal in the Beresford Restaurant around the corner from the gallery before returning to the hotel. A table in a secluded corner had been booked. It was a relief to get away from the hectic hand shaking and congratulations. As they sat studying menus each was overwhelmed at what had taken place. Adrian Osborne had whispered that eighty-five per cent of the works had been purchased at the private view and the show was scheduled to go on for a further two weeks.

When their orders were taken Tom beamed at his guests.

'Both of you have made a great impact. I must say John your speech was a masterpiece. I never expected a painting of mine to be compared to the human body -- and mind for that matter. Never heard that before and especially coming from you. It really meant a lot. It made terrific sense and everyone understood where you were coming from.'

'I couldn't speak as an art critic,' he said easily. 'They were probably present just waiting for me to fall into some politically nasty trap and say the wrong thing. Although I was tempted to mention things like making a statue of the Madonna out of elephant dung that won the Turner prize. And there have been worse examples.'

'Enough said on that subject John,' Tom replied contemptuously. 'I go apoplectic listening to the goings on about the Turner prize, at IMMA and the Arts Council. Now I'd like to propose a toast to you and Karita. Thank you for being you -- true to your principles and for taking the trouble to travel to Dublin. So a double thanks to you, John, for tonight and also for giving us back our daughter, Annette. You'll be pleased to hear she's almost top of her class in all grades and her teachers are so pleased with her.'

At the end of the meal Karita spoke directly to Tom. She waited for an opportunity.

'Tom before we break up there is something I've been dying to talk to you about.'

'What's on your mind?'

She turned to John. 'Hope this is all right with you.'

'Go ahead Karita. I don't mind.' Yet he wondered what was coming.

'Well Tom. It's like this. I've written a Trilogy on Celtic Mythology; the first two books have been published and well received. I've been working night and day on the final section and have come across several blank areas. When I saw your brochure I was overwhelmed and thought this could be an answer to my prayers. What I have in mind is a collaborative study in the final section between the written word and some illustrations. What I mean is to include artistic representations of the important places in Celtic Mythology for example Dun Angus Fort in Aran, Stague Fort in Kerry, the Skelligs, New Grange along with some others. If you would agree to cooperate then I would approach my agent and see if it is feasible. Everything is in the air at the moment. Would you be agreeable we should go ahead with the proposal?'

'Yes of course. I'd be delighted Karita.'

*

Some days later as the grey fog pressed against the cottage windows in the little back garden, Tom, Ruth and Annette stood together in a pathetic group. Tom had just hammered a plaque into the soft ground. On it was written; 'Beneath this earth lies the remains of our dear friend, Patsie, who had beauty without pride, strength without contempt, courage without cruelty, and all the virtues of humans, without their vices.'

Tears streamed down Annette's cheeks and she let them run. Tom saw this and gave her a big hug.

'You know, Annette, I've a terrible memory for forgetting and sometimes it helps. I think you will learn that it is much better to forget and smile, than remember and be sad.'

She looked at him.

'I'll never forget Patsie.'

# 20

That Sunday morning the sky was clear after last night's rain and it was warm enough to sit on the terrace of the cottage. The air was full of the fragrance of fresh blossom and damp earth – this is where John felt closest to the heart of his creator. Karita wished to discuss over coffee the interview with her agent Liz Duckworth. She'd a luncheon appointment in London the following day.

'What are you going to say to her?' John asked casually.

'I'll tell her part three is almost finished but there are certain problems, which I'm finding hard to complete.'

'Difficulties?'

'Yes. Concerning places, which are difficult to describe in words and after Tom Nolan's exhibition and brochure, I thought how wonderful it would be to include some dramatic illustrations to complement the text.'

'Sounds logical. Liz will probably ask why not photographs? Wouldn't they be just as good or even better?' John asked with mischief.

'Oh never, John. She was shocked. 'I can see you don't really mean it.'

'No, but others will. I'm merely the devil's advocate. So what will you say?'

'That'll be hard. Perhaps you can help me formulate my thoughts. Your speech, which was brilliant, stated that creative art is not necessarily what we see with our eyes. It's more and has to be if it is going to work. I like that especially when you combined Tom's words about searching for mindscapes that no one has seen in landscapes everyone has seen.'

'Still doesn't entirely answer the question of photographs?' John said provocatively.

'A photograph is objective and impersonal, you said. It's like a mirror that reflects what it sees with no preconceptions, no prejudices, no qualifications and no emphasis. No opinion at all. It's stone dead.'

'I remember that. I just got a bit carried away.'

'But it's true! They're lifeless whereas Tom Nolan's paintings have a unique and private vision of the world. He's not merely describing the visible appearance of a landscape, but telling us something beyond itself. Often it's an original discovery by the artist, which he has to communicate. The greater the originality the greater the work and he certainly has the technical skill to carry it out.'

He frowned slightly. 'Know what you're driving at. By a close observation of Nature he's able to discover qualities I'd never have thought possible or ever seen with my own eyes.'

She gave a sigh. 'You know something, John. I wish you were coming with me tomorrow. Liz Duckworth is a tough cookie and it's going to be hard work trying to convince her that this is the way forward.'

Oh Lord. Not possible, thought John. So much going on in the hospital and so many pressures building up over the UGC report. If it were any other time – well maybe. Better for Karita to fight her own corner anyway.

'Sorry. Can't be there. Too many problems in the hospital that need attention,' he smiled confidently. 'But Karita you're the one who has a wonderful way with words. Sometimes I think you could convince an artist that black is white.'

'You say the nicest things, John. That's encouraging. If it works out it will be wonderful. If not I shall have to live with it.'

'One thing you haven't considered. If it does work out and the publishers accept your proposal you're going to give Tom Nolan a large stage on the international market because your books are destined to be a big success. We've been told that already.'

She said no more and excused herself. More notes and writings had to be done in her room before tomorrow morning.

Later the morning grew warmer, scents moved seductively through the air and as the sun came out from behind a curtain of cloud John did a little gardening -- his favourite relaxation. Amazing what a few days neglect could do to the appearance of a garden.

After the evening meal they sat in the living room. He enjoyed nothing better than hear Karita reading a new piece of her work. As the day grew older tiredness descended on both. Life felt good in their temporary security yet they still had to go separate ways tomorrow.

And John Nicholson had worries about Karita's proposal, which she was determined to carry through.

Returning to her room she felt she should prepare a document for Liz Duckworth to help formulate ideas in support of the proposal. There would be difficulties in the editorial committee and any additional ideas would probably be welcomed by Liz.

She started to write.

... For thousands of years primitive races have felt the need to express themselves in some form of realistic art to illustrate what they saw in the world around them; many famous examples remain with us to this day. The best of them can put us to shame when compared to some public art on display ...

... In spite of this if any comment is to have a purpose to it, it must be to convey the idea behind the work, the reason for it being produced. Normally that task is left in the hands of critics and experts ...

... The main opponent is time – more precisely the lack of it. In music we can hold the entire score of a great work in our hands, yet it will not convey the essential idea of the composer to us. We must hear it over a period of time. When finished we then reflect on what we've heard and attempt to express our reaction, to interpret its sublime effects on our emotions and analyse its quintessence ...

... So too with visual art. We see it – for a time – and we like or dislike it. We should be able to share with the artist the complex procedures that went into the emotional states the work attempts to convey and then as a writer it is my task to uphold or reject the ideas behind the work. Sometimes words are not necessary. The work speaks for itself. To me this is especially clear in Nolan's work and generally needs little explanation. Most people love it. Because our lives are full of sweeping inclusion and sometimes confusion his art is full of acute discernment and selection, which makes it so easy to interpret. Contrary to many forms of 'modern art' where imagination without skill is rampant and generally praised in words no one can really understand ...

... When we attempt to write about art – and in site of spirited denials – the professional critic is in the same boat as the proverbial man-in-the-street. We begin an almost impossible task. Art begins where words end. It is the outward and visual representation of an inward and personal reality, suggesting there is something else beyond and behind what we see. Some would argue this is not the case. Is it not the art in front of us that makes the thought, or the symbol that makes the unique reality? Art begins where words end ...

... Still this leaves many of us in the dark. Many are speechless. Experts usually try to get around it by avoiding the simple and profound problems by taking refuge in technicalities that the ordinary person finds confusing, even bewildering. As a result the most natural works are apt to be the most embarrassing for the expert critic who has taken these problems as solved. He has no common sense answer but gibberish and evasion. The man-in-the-street is perpetually kept in the position of a child who has not grown up or does not possess the intelligence needed to understand this gibberish ...

... It is hoped that something can be done to break down this barrier of imposed modesty of the plain man by showing him that art is not confined to the esoteric land of the exclusive intelligentsia who attempt to impress lesser mortals with pseudo-critical jargon. The only honourable thing for them is to educate and enlighten the public that in the end he himself will become redundant ...

... If we like what we see we see what we like. And Tom Nolan does not reproduce things that are visible. He makes them visible.

Karita Isselherg.

# 21

**K**arita Isselherg was getting used to the train journey to King's Cross. Her briefcase was full --perhaps too full -- yet insecurity dictated she required all the evidence she could muster and was determined to get her own way.

Liz Duckworth was particularly pleasant as she welcomed Karita into the office. Coffee and biscuits were provided and informal chat filled the air as if they had all the time in the world. Suddenly Liz put her cup down and stared at her.

'I hear you wish to change the final, the most precious part of The Trilogy. How could you stray off the beaten track Karita? This was to be an erudite account of Celtic Mythology in Ireland in prehistoric times -- a more or less fait accompli and now I get a last minute SOS. Please could some painter's work be included in a purely literary account? Quite frankly, Karita, I don't understand and I cannot approve it. It goes against the grain of the other two books. Can I be frank? Have you fallen for this guy? What's his name -- Tom Nolan? Or is there more to it? You know it'll take a lot of convincing. I've been through last minute changes with other authors and it can cause havoc in detailed planning of the publication procedure. It's throwing a spanner in the works.'

'Sorry you feel that way Liz,' she said in a soft courteous voice.

'Remember I've had more time than you to consider the merits of the proposal. If you mention photographs as a compromise, they're out. They're dead meat. Cannot be used. This Tom Nolan character has the power to provoke a whole world of ideas, which is original and has caused amazement.'

'I've seen the brochure you sent me. It's remarkable and the writing's true to the mark although some refinement is required. But Karita dear, do you have to go through with it? It'll upset an almost completely finished contract and bring in extra complications. It's too bothersome.'

Karita leant forward pleading. 'Liz. I desperately want this to go through even if it kills me. I've put so much work into the whole project I want this to be the grand finale, something that people will say as The Trilogy developed it got better and better.'

'You know, Karita, I have strong views on books and writing – I should know. I think you have the gift. All impressive books I've read seem truer than life, and when I have finished I'm filled with a sense of loss. I just *feel* all that has happened to me and afterwards belong to me; the joy and sadness, the corruption and solutions, the ecstasy, sadness and sorrow in places and even how the climate behaved itself. If you can give that to your writing, and I believe you can, then you are a writer. I know you write so clearly therefore you'll have readers – but if you write with obscuration in mind you'd be lucky to be left with mere critics or none – none at all.'

'Thank you for your comments, Liz. I find writing a lonely life. I work alone – have to – and if it starts well I encounter comfort, eternity, or opposite, each day

'OK. Karita. I'll have to put it to the editorial committee and it could take some time. I suppose you can continue to liase with your artist friend, Tom Nolan, and go ahead with your commissions. Wish me luck!'

# 22

Sir Kenneth asked to see John Nicholson as a matter of urgency. The University Grants Committee had written to the medical Deans of UK universities requesting details of current and proposed research. This was normal practice and occurred at regular intervals. What concerned Sir Kenneth was the reference to the Oncology Unit in Addenbrooke's where there was some question of irregularities and more information was sought.

'You must understand John this is pretty routine stuff.' He tried to sound reassuring. 'Everyone's required to submit a progress report and Addenbrooke's is no exception. Yours is just one of many.'

'Why did the Q E 2 in Welwyn come up?' John asked frowning slightly.

'Someone probably alerted them to your protocol with Wilkinson. These things get around in mysterious ways,' he continued, indicating with his eyes, the room, the surroundings.

'And ... '

'And they want to know more about what's happening,' he said quickly looking at John.

John replied modestly. 'I'll be glad to let you have a report giving details of our procedures. It'll be technical and complicated and may take time.'

'I realise that. To make things easy could you give me a short summary of procedures when a cancer patient is referred to you? I can compare your approach with others.'

John gathered several sheets of paper lying on the table. He drew diagrams and chemical equations.

'All routine diagnostic procedures are carried out when the patient is first seen including radiology, ultrasound, CT scans and haematology,' John said in a soft voice. 'Biopsy is of particular importance and that's where our protocol begins to differ from others. We biopsy the cancerous lesion and then take another 'clean' biopsy of normal surrounding tissue. Half the cancer biopsy is sent for routine report and grading. The remainder is stored in liquid nitrogen for tissue culture lines. The 'clean' is also stored and used in culture lines.'

Sir Kenneth nodded as he intervened. 'I see. Go on.'

'The cultures are then subjected to various agents in different concentrations and with specialised equipment. Cellular responses are measured and recorded. The slightest change in behaviour of normal and abnormal cells are noted and gives us an opportunity to almost tailor-make specific drugs, or more often, a cocktail of drugs that will have maximum effect on cancer cells and minimum damage on normal ones.'

Sir Kenneth smiled and looked at him with a warm glance.

'Sounds like the closest you could get to a dress rehearsal of the chemotherapy process in the actual patient ...'

'Without the tedious practice of obtaining mass-produced agents from animal testing, when it's assumed if there is no significant damage to them then the agents might be safe to use in humans. The company asks for a licence to mass-produce the one product, which will probably be used in different cancers in a wide range of patients, young and old. To me, this is blunderbuss or hit-and-miss therapy. Sometimes it may work.'

'Steady on John,' Sir Kenneth replied a little breathlessly. 'Those are hard hitting words against widely carried out practices. It's what most cancer patients are subjected to.'

'Yes. Some are lucky. Lots more are not.'

'I see what you're saying. It's obvious that you, John, are the lucky one because of the experts who work with such highly sophisticated equipment and the ability to modify chemical agents to suit individual patients. Not every hospital in the country can hope to have such an array of advantages and even the know-how.'

'You're right and it comes down to funding,' John drawled, thinking deeply. 'That's probably why the University Grants Committee is interested in how our system works and more importantly how much it costs. I can see accountants doing a cost benefit analysis of each treatment. I wonder what kind of tale of woe they'll dream up for the Department not thinking of the billions of profits being made for pharmaceutical companies with their mass-produced products used in the NHS.'

Sir Kenneth laughed evasively. 'John you've summarised your approach well but you have given me a difficult task in convincing the Committee your way is right and the rest is wrong. I'll have to be more diplomatic.'

John answered flatly. 'I suppose you must to keep the peace.'

'One last thing and it is important, carry on with what you're doing but always make sure you get the patient's written consent before proceeding otherwise lawyers may be after us and that could mean trouble.'

*

Karita was waiting when John arrived home. He'd to hear details of the interview with Liz Duckworth. Liz was reluctant to go along with the proposal about Tom's illustrations. It would have to be submitted to the editorial committee. She was determined to go ahead with commissioning the work. She'd already spoken to him by phone and it was suggested they identify what locations, what backgrounds she'd in mind, what to show as well as leave out. She'd take the manuscripts with her and work on location.

John's reaction was sudden and bitter. She'd be away with a

relative stranger for two or three weeks -- an added burden and a worry about her well-being. Was she risking too much for the sake of her writing, risking perhaps a rift in their relationship? He could hardly object although was tempted. Time spent apart from Karita was becoming a void. She didn't realise it, or purposely made light of it. After all three weeks, she argued, was a short time when a lot could be achieved. In addition Karita's location tour was scheduled to take place during one of the most critical times of John's career as he was faced with writing a detailed report, which was probably more important than Sir Kenneth had implied.

John also suspected certain forces were anxious to isolate what he was trying to do and, if possible, prevent progress. His report must be as foolproof and watertight as he could make it. It was up to him but what a task! He seemed to be the only one pursuing the in vitro technique. It was a lonely business.

Fate sometimes selects strange circumstances for conflict; the next few weeks looked like one of these periods. He tried to console himself and not think of the future. It will happen soon enough even though it only comes one day at a time.

# PART TWO

# 1

As John Nicholson concentrated on the university report he asked not to be disturbed. Full of apologies Gillian, his secretary, phoned to say a Dr. Wilkinson from the Q E 2 Hospital wished to speak to him urgently. This could be relevant.

'Nice to hear from you again, James,' John said quickly, 'what's on your mind?'

'John,' he said emphatically. 'You won't like me say this but we've come to the end of our tether with another patient. She's pleasant but difficult to manage.'

John listened carefully. 'Go on.'

'She was referred four weeks ago with a reticulosis -- a non-Hodgkin's lymphoma. The diagnosis was confirmed and conventional treatment started to which most cases improve, as you know. Not Karen Smith; although she made an initial response, after two weeks she went downhill. We increased chemo and auxiliary treatment for another two weeks to no avail and now she's poorly. So much so I fear for her life.'

'Let me guess,' John said with infinite patience. 'You'd like us to take over.'

'How right you are. She's an exceptional case, a lovely young woman with an intractable problem. This is a cry for help. I should also tell you there's a difficult family background. We cannot solve it here but if you agree to accept her you may have better luck.'

'We'll see.' Alarm bells rang in the back of his head. More trouble!

'I appreciate if you could take over. It isn't straightforward. There's only one unit in the country that can save her and it must be yours. If she stayed here no doubt we'll lose her.'

Karen Smith arrived in Addenbrooke's within twenty-four hours and John Nicholson was dragged away from his paper work to interview her. Warned of pitfalls he was extra careful.

He sat at her bedside and smiled.

'Karen Smith, welcome to Addenbrooke's. I'm Dr. John Nicholson and I hope you'll make progress with us. I've had a long talk with Dr. Wilkinson.'

'Sorry Dr. Nicholson.' She was distant at first. 'I don't understand why I have to be transferred to another hospital to have the same treatment. It doesn't make sense.'

'Did Dr. Wilkinson not explain why you were being transferred?'

'Not really. Although he said you'd better equipment and some form of new treatment. My husband would think that unacceptable. He'd say the Q E 2 should have exactly the same facilities as you have here. And there should be no need for him to travel forty miles extra to visit me. And Dr. Nicholson I agree with him. I'd prefer to be in the Q E 2 receiving treatment there. It's much closer to home and my friends can easily visit me, whereas now I'm miles away. It doesn't seem fair and I don't like it.'

John relaxed in his chair and attempted another smile.

'I'm sorry Mrs. Smith,' he replied modestly. 'You appear to have missed the point entirely. Addenbrooke's is a highly specialised hospital in the treatment of conditions like yours. We are pioneering techniques that are not generally available in other parts of the country.'

She sat erect and glared at him. 'Why not, Dr. Nicholson?'

'That question is at present being thoroughly investigated by the University Grants Committee and there's little more I can or should say. The treatment you will get here is at present unavailable elsewhere but is proven to be scientifically correct.'

'Why is it unavailable elsewhere?' She persisted.

'Because it's a new technique whereby we test your normal and cancer cells in the laboratory, each is subjected to variable drug doses and with the equipment here we can make fine modifications to the drugs, which are then given to the patient.'

She was beginning to relax. 'You make me feel in a privileged position.'

'I suppose in a way you are. There's only one requirement before we can go ahead and that is the hospital needs your written consent.'

Her smile disappeared immediately. 'Oh but Dr. Nicholson. I cannot do that. My husband has always advised me never to give written consent to anything unless he signs it himself.'

'There should be no problem. We'll find your husband and ask him to give consent.'

She shrugged. 'Oh dear. That could be difficult.'

'Why should that be Mrs. Smith?'

There was no answer.

'Mrs. Smith, I repeat why should it be difficult to find your husband?'

She looked away. 'He's at a conference.'

'Where is the conference?'

'In Bournemouth, I think.'

'Mrs. Smith. You must tell us how we can find your husband. What is his profession?'

She hesitated. 'He's a journalist.'

'And ... '

'He said he'd be away for a week in Bournemouth but I don't know exactly where.'

'You realise the gravity of the situation.' John stood up and faced her. 'We must contact him as soon as possible. You're not well and he needs to be told.'

'I don't know about that, Dr. Nicholson.'

'Why?'

'He's a difficult man. Always has been and sometimes we don't see eye to eye.'

'Well then. All we need is your consent. Will you do that -- for yourself, if not for your husband?'

'It's difficult for me doctor.' She was a little shocked. 'You must understand. It's very difficult. I cannot decide.'

She started to cry.

'All right.' John said reassuringly. 'We'll leave it for a while but we cannot put it off indefinitely. Think about it for a couple of days.'

'Very good. But the answer will be the same. My husband would kill me if I did anything without his permission.'

'Does he not phone you when he's away?' John asked a little surprised.

'Not very often. Only if he wants something.'

John smiled and turned to go. 'All right. I won't disturb you anymore. You relax and leave things to us. I'm sorry to have upset you Mrs. Smith.'

He nodded to the ward sister and left.

The search for Peter Smith started immediately. John Nicholson phoned Chief Inspector Paul Jordan in the local constabulary and explained the situation, emphasising the urgency of finding the husband.

'We've to get a decision out of him quickly. Apparently he's at a journalists' conference in Bournemouth at present and doesn't communicate with his wife much. He probably thinks she's still in the Q E 2 Hospital in Welwyn.'

'How sad,' Jordan replied with a sigh. 'Not very caring, is he? Don't worry Dr. Nicholson we'll get one of our men on the job and ask him to contact Addenbrooke's. That's if we can find him.'

'I hope you can.' John was emphatic. 'It could be serious for the patient and time is running out.'

'We'll do our best. I'll get back to you one way or another.'

'I'd appreciate that.'

John thought flippantly about people killing time, but now it was rapidly killing Karen Smith.

# 2

Next morning Sir Kenneth Richardson sat in his office behind the impressive bulk of a desk surrounded by papers, journals and correspondence. He had slept well, had a light breakfast and his fresh complexion was clean and shiny from the razor. He was feeling good-humoured and welcomed the busy day's work ahead. In the large room, with its elegant stucco-worked ceiling, thick carpets and impressive portraits in guilt frames, he was endowed with an obvious dignity of possession and position.

Today, however, there were serious problems to be faced. He was well able for the highly-strung temperaments shown by some high flyers on his staff; that and intelligence often went together and sometimes had to be handled with kid-gloves. Any university – Sir Kenneth sighed – was a hot bed of animosities and jealousies, often over petty issues.

But not so with John Nicholson. He sat in front of him pale and worn and his clothes hung loosely on him. His eyes were sombre with fatigue and deep furrows were evident on his brow. His voice was flat yet urgent.

Sir Kenneth had to be told.

'It's a damned nuisance the way some people behave,' he said angrily when John finished the details of the Smith case. 'Going off like that and leaving the poor wife. She has my sympathy.'

'She needs more than sympathy, Sir Kenneth,' John said but regretted it. 'She needs proper treatment.'

'And we need her consent. Without it we cannot go ahead.'

A feeling of déjà vu confronted John. He was up against a brick wall -- again.

He said almost pleading. 'Is there no way out of the impasse?'

Sir Kenneth's face darkened as if he was consumed by some inner struggle. John watched him anxiously, afraid of having upset him. After a few moments Sir Kenneth looked straight at him his face still serious, but his eyes were filled with gentleness. He said in measured tones.

'I see no way. I can only be guided by rules and regulations. I'll say no more than that John. They leave me no option. Must abide by them.'

'We'll either have to persuade the patient or the husband. If not then ... '

'Then continue with routine chemotherapy.'

John nodded clenching his fists until the knuckles whitened.

'Thank you, Sir Kenneth. We'll just have to wait.'

'Yes, John. We'll wait and see.'

It was a rare moment of aloneness for John. Others seemed not to care, or were so obstinate. It defied belief. Some people were strong enough to bear the misfortune of others – not so John Nicholson. He was devastated by the inertia and inaction.

*

Karen Smith's condition deteriorated over the next two days. Confusion and disorientation set in. Her medication was changed but still had little effect. John Nicholson was a frequent visitor to her bedside and tried to reason with her. It was difficult.

On the third morning he spoke gently to her with ward sister in attendance. Karen had difficulty in breathing and was in pain.

'Karen, Dr. Nicholson here,' he almost whispered.

'Who?' She asked in a confused state.

'Sister and Dr. Nicholson. Can you hear me?'

'Yes. I can. I do feel awful.'

'We know. And we are worried about you.'

She closed her eyes and remained silent.

'Karen. We want to help. You must let us help you.'

'I feel so weak.' Her eyes were still closed.

'We know all too well Karen. Isn't that right sister?'

She nodded her head. 'Yes doctor. We really do.'

'So Karen. I ask you again, can we go ahead with the new treatment?'

'What treatment?'

'I've asked you before and explained everything. Don't you remember?'

Her eyes remained closed. 'I remember now. The new treatment. I remember.'

'That's good. Will you give us permission to use it?'

'Yes. I will. Please go head.' She almost pleaded. 'Anything to help me. I feel so awful.'

'That's good news Karen. We'll start treatment as soon as possible.'

'Thanks doctor. But I can't sign any paper. I just cannot. I'm terrified to do so.'

'Don't worry Karen. That's not important. Sister and I know you've given permission. Isn't that right sister?'

'Yes. Dr. Nicholson. I've been listening carefully.'

Karen Smith kept her eyes closed and appeared to lapse into a light sleep. John went to the laboratory and set in motion new treatment protocols thinking that desperate means were required in desperate situations.

Chief Inspector Paul Jordan telephoned John Nicholson the same day informing him that there was nobody by the name of Peter Smith in attendance at the National Union of Journalists' meeting in Bournemouth. Also there was nobody at the Smith's home address in Welwyn Garden City. So for the present the husband of Karen Smith could not be found. However efforts would still be made to trace the said person. They realised the importance of the situation.

Karen Smith's condition improved over the following days, her breathing became easier and she was in less pain. She was more alert and could talk for longer periods with staff.

John Nicholson was much relieved and continued to increase the dose of the in vitro medication and measure the response.

Things were beginning to look up.

*

As they raced through the countryside in Peter Smith's open top MG Penelope closed her eyes and rested her head on the neck restraint. The cool breeze rushed around her ecstatically and rippled her long golden hair. Peter concentrated on the twisting roads and had little time to admire the passing landscape. They were supposed to be going to Bournemouth, but actually heading for Hastings – not too far down the coast.

As they passed tangled green coppices tall trees spread out to embrace each other across the roads. He wanted to arrive quickly; she just enjoyed the journey. Occasionally they came across people in fields working on their crops – they were like intangible shadows lent by the sun to the earth, not really for toil, but for some long established tradition in the golden autumn air. Over the whole scene of tall trees turning yellow and red, some shelters mainly for cattle, and lazy streams that took short cuts across the side roads, drifted the gentle heat, never threatening, only soporific if one let it, and was like a great warm mother figure nourishing the grateful earth.

Days later in the Broadway Hotel in Hastings Peter Smith settled his account.

'I hope you and Mrs. Smith enjoyed your week with us sir,' the receptionist said at the checkout desk.

'Yes indeed. We'd a pleasant time. Perhaps we'll come again some day.'

'That would be lovely, sir. Safe journey home and thank you both for your custom.'

She gently shook her head as they left the front door.

# 3

Graham Sinclair was in plenty of time to meet the incoming flight from Rome. Helen Watson was returning from a week's holiday and he'd promised to be there when she arrived. As she walked towards the exit she saw him waiting patiently but was shocked by his appearance. And yet as soon as she jumped into his arms with the joy of meeting she forgot about it.

'Graham. Graham. So lovely to see you.'

'And you Helen. Had a good time?'

'Yes. Sensational and exciting but it's great to be back. Have you missed me?'

There was a short silence as she noticed new lines on his brow and instead of answering he fussed and fretted about her luggage.

'I'll have to get back to the hospital as soon as possible.'

'Why is that? You usually take things in your stride.'

There was no answer. So they made their way to the exit.

Later when they were alone she asked.

'Why are you so pale? Is there something wrong? Now that I'm home again everything will be all right. We'll have each other properly in the near future. So don't worry.'

He looked away and put his head down.

'You're not in any trouble, Graham?'

'No. Of course not. It's just that...'

He shrugged and looked around in a furtive way. He shook his head.

'You don't understand, Helen. Things have changed.'

Again his eyes were staring at some fixed point in front of him. She touched his face and made him look at her.

'Again Graham, you're not in any trouble. Are you?'

'No. No trouble.' Again his eyes were fixed on some point.

Suddenly Helen changed the subject hoping to get around the impasse.

'Jane and I saw an enormous amount in four days,' she said enthusiastically. 'Those CIT buses are marvellous and some of the places are just unforgettable.'

'Great. Great. Good for you.' He was still looking away. 'You must show me your photographs sometime.'

'Love to. When they're all processed.'

She started to unpack and casually asked again.

'How are things at the hospital?'

'Oh, the same old routine – although the work gets harder and therefore shoddier because of all these new targets brought in.'

She was puzzled.

'Targets. I don't understand. I thought you were supposed to be treating patients – not playing bows and arrows.' She laughed.

'Helen it's not funny. It's infuriating. The harder we work the greater the work load and we're not allowed treat patients anymore. Now we have clients to work on and discharge as soon as possible. It's like working in a motor mechanic's garage.'

She looked carefully at him. He was handsome and his face so kind. Never in the six months of their engagement had she seen him this worried or dejected. Before she left he was confident and well adjusted. What on earth had gone wrong in the meantime?

'Graham, please tell me quietly what has happened,' she pleaded looking straight into his eyes. He tried to avoid her but she wouldn't let him. 'I love you dearly. You know that and perhaps there is something I can do to help. I can see something has happened, probably in the hospital and made you so gloomy. Telling me will probably do you good. Please.'

Graham embraced her tightly. 'You're a lovely person, Helen, and I love you dearly. There's nothing wrong with my health – according to a recent check-up and everything's OK. I was so glad because of the possible changes in the future.'

'Then we'll be married in three months?'

He was silent for a moment and then said weakly. 'If you think that would work out financially.'

She laughed a little nervously. 'I don't understand. What do you mean? You've a good job and great prospects according to Dr. Wilkinson. With time you will probably get a consultant job.'

'And end up like him. No thank you.'

At this stage she was getting a little angry and said quickly. 'What more could you ask after all your hard work? Most consultants are happy and reasonably well paid and...'

He almost shouted at her. '...And frustrated as hell.'

'Shut up Graham. I will not hear any more of this nonsense. You've got a bee in your bonnet about something. Perhaps a row with a colleague or administrator. These things blow over. They usually do and everything is back to normal in no time. I can't stand idly by and see you fretting over some silly misunderstanding at work – and now behaving like a spoilt child.'

'Helen, I think of you and our future when I resign my job from the Q E 2 in three months time.'

Helen began to recognise she was looking at a sick man and, for the moment, there was not much more she could do or say. She tried to calm him down and listen in a semi-conscious way about ambition, the present and what prospects the future could hold.

She felt something implode inside her. Her hopes, joys and happiness dashed without argument and it didn't make any sense. She even suggested they go away together for a break to work things out.

As time went by she found that her tenderness and affection were not enthusiastically received as before.

Perhaps there was another woman. She was frightened.

A few days later Graham came to see her with red rimmed eyes and fatigue lines on his face. He spoke hesitatingly.

'Helen, you may think I'm crazy,' he said in a staccato voice. 'You could just be right. There are times -- especially in the morning when I wake up – when the problems seem to lift and I feel better about things. But unfortunately those periods are getting shorter. I even think Dr. Wilkinson feels I'm failing in my work and he has had to rebuke me several times over stupid mistakes. I guess that if this continues I'll be asked to leave.'

He talked logically and calmly. He wanted to at least postpone their planned marriage when he felt their engagement hang in the balance. But for love of him she was determined it was not over and her hope was still active and alive and her actions had already begun.

# 4

The next day she made an appointment to see Dr. Wilkinson – Graham's direct senior.

She immediately came to the point.

'Have you noticed anything odd about Graham's behaviour recently, doctor?'

'Yes indeed. He's doing some strange things and goes around as if he's plotting to have a nervous breakdown. I've noticed some mistakes and these could be serious so I've had to tell him off several times.'

'All this whining at silly things and looking to blame others. There must be some reasonable answer.'

'And what do you think the reason is Helen?'

'I don't know but goodness I have tried hard enough to find out. Perhaps, I could ask you a big favour. You have a personal chat with him to see what's on his mind. Would you do that for me at least doctor?'

'He'll have to find that out for himself, Helen. I can only go so far and listen to general questions but perhaps it's worth a try. He really does have good potential. It would be a pity to see it wasted.'

'Will you let me know how you get on?'

'Yes. But no promises. I do not have a magic wand.'

*

Two days later Helen got a phone call from Dr. Wilkinson who asked her to come and see him at her convenience. An appointment was made and she arrived at his rooms in good time.

'Please sit down Helen and make yourself comfortable.'

He rang for tea and biscuits to be sent in. 'I've had a long chat with Graham. It was difficult at first. He was full of complaints about the new rules and regulations and the load of unnecessary paper work that seems to be increasing like a snowball going downhill. I sympathize with him, of course, and said we all are now facing a new system called reorganisation of The NHS – to make us all more efficient.'

He stopped for a moment and looked into his teacup. 'You know Helen, I really don't think that was the main problem at all. It's only a minor irritation. I had a feeling there was something far deeper eating away inside him.'

'So you persisted?' She leant forward and listened carefully.

'Yes. Eventually it came out that he was unhappy with the type of work he was doing – routine surgery – and fulfilling his time for the FRCS – when he could and probably would make a jolly good consultant.'

'Well. Did he not agree?'

'No. Most certainly not. It was at this stage he got up from his chair and started pacing up and down the room. You know he worked in Addenbrooke's in Cambridge before coming here?'

'Yes. And his family still live there, I think.'

'That's right. But it's not for family reasons he's upset. It appears for sometime he worked closely with Dr. Nicholson and was extremely impressed with the whole set up. He even asked Dr. Nicholson would there be any chance of him getting a position in the Oncology Department.'

'And?'

'John said unfortunately no. Not at present. Funding was tight and graduates were needed who excelled in information technology plus medicine. In other words Graham would have to go back to university for a couple of years and do, at least, a Master's Degree in I.T. if he were to be considered at any time in the future. So poor Graham has a dilemma – continue on a surgical pathway to a safe but dull career – or go back to college to take a chance on doing something he would really be fulfilled in.'

Helen lowered her head and felt she had been given a blow in the solar plexus.

'And that dilemma involves me – big time. We were due to be married in three months and settle down to an ordinary routine of married bliss. Now it appears his ambition has grown out of all proportion and he can see no way that he can do both things.'

'I think you've got it in one. He loves you, I know that for sure. That's why he is so upset because if he enrols for a Master's Degree course on computer technology financially it will be a noose around his neck. Humans cannot tolerate too much reality – especially all in one go.'

'But we could survive. I have a reasonable job and he could do locums in the evenings, weekends and holidays.'

'That's brave talk Helen and I admire it. But this new pathway is so uncertain. Even if he completes the course and does well in his examinations there's no guarantee of a good job at the end of it all.'

'Remember The Bard of Stratford said security is mortal's greatest enemy or something like that.'

'I remember. And nothing ventured nothing gained is an old cliché.'

'Dr. Wilkinson, I'm glad you were there to drag this out of him. I don't think I could have done it. But now I know I can plan a way. There is bound to be a solution.'

'I've already said Helen you're a brave girl but you probably don't realise all the hardships ahead.'

'Look, Dr. Wilkinson. I love him. He's the only one for me. If he does not do this and takes a chance on it he is going to regret it for the rest of his life – and also make my life completely miserable too.'

Dr. Wilkinson stood up and shook her hand. She certainly was a determined young lady.

# 5

Even through it was early morning in Stockholm Karita Isselherg lit three candles on her desk in the library of Kungsangan House. She preferred to be at home at this time -- less stressful, more peaceful and she could get on with writing. In Cambridge John had become difficult for the past few weeks, moody, on edge most of the time and usually brought work home in the evenings. He'd little time for her; his mind elsewhere. She'd tried to calm him and make matters more comfortable but to no avail. Things went from bad to worse and she decided to return home. It was best for both. The only way. Now her work was easier, flowed better and ideas came without obstruction or interference. Word from Liz Duckworth was eagerly awaited.

As time dragged by it became more acute. Yet it helped to take her mind off John and his problems. Before leaving London she informed Liz she'd be based at home for a few weeks. For several days she worked on complex passages, made more difficult by a lack of a decision from the editorial committee. If news was positive most of these would be resolved.

The postman arrived, but was empty handed; the phone was silent except for inconsequential messages from friends inviting her to parties because they knew she was in town. They held little attraction as she carried two unresolved burdens, Liz's decision and John's intransigence.

Several ways of distraction were open; her favourite was to walk down to the river with its silky reflections, changing and moving from her special vantage point on the bridge. Today the air was clear and warm, birds and flowers were excited about the approaching spring. For hours she'd sit on the bridge, daydreaming and completely undisturbed except by flashes of inspiration that were jotted down immediately.

It was a busy time, as the deadline loomed. Parents and family tried to help and sympathize but she missed John. The phone was some help and used frequently but it wasn't the same as sitting directly, talking silly intimate things, laughing, joking and being happy. He was still going through some secret rough patch and she prayed it would blow over soon. It was becoming intolerable.

One morning a large envelope arrived from London addressed to Karita Isselherg. She ran into the library, with heart pounding and opened it cleanly with the ebony letter knife.

It was from Liz Duckworth.

'Dear Karita,

Re Celtic Mythology--A Trilogy

I had a long and sometimes difficult meeting with our publishers yesterday. Many items were on the agenda and discussed at length and some heated arguments took place.

After about an hour the chairman invited me to explain your proposal to include a selection of Nolan's paintings in your Trilogy. I did my best based on your promptings – boy, wasn't I glad we had lunch together to thrash it out! The chairman seemed impressed especially when he saw the brochure. But there was an irritating little man who objected strongly saying they'd be biased, not truthful, even childish, and might cheapen the end product -- like a comic strip of all things! It didn't make sense until he said they'd be too expensive. Photographs would be more believable and it's what people expect in magazines today.

I was incensed at this gibberish. He seemed to have lost the point of your manuscript. In fact I'd a sneaking suspicion he hadn't even read it the way he was talking. I tried to put him straight saying this was a unique work of imagination, a writer's imagination and now a brilliant artist's imagination was needed to complement the writing. Why spoil it with dead photographs?

The proposal was put to a vote and was carried by a majority; you'll be glad to hear. Even though the chairman was pleased he did say the publishers would only be prepared to give a grant of a further £10,000.00 for you and Nolan for the necessary location work. If it goes above this you're on your own, Karita.

The result is favourable and the one you've been looking for. So congratulation on what I feel is a fairly major achievement.

I trust this news will make the completion of your wonderful work more exciting and fulfilling. So it's over to you to make the necessary arrangements as you see fit.

Hope all is going well and look forward to hearing from you. Incidentally do not hesitate to contact me if you need further information or assistance.

Always glad to help,
Every good wish,
Sincerely,
Liz Duckworth.'

*

Karita phoned John later that evening. He was delighted. It was what she desperately needed all along. A big obstacle had been removed and the way was clear to meet the deadline and to commission Tom Nolan -- although that business of location work was indefinite. What exactly did it mean?

As he put down the receiver he was overcome with emotion. Yes, she'd the news she was waiting for. It would improve the quality of the work and probably place it in a unique bracket. She deserved it; intelligent, tenacious and forceful in getting what she wanted.

One worry remained -- the location work with Tom Nolan. This meant travelling to historical places and knowing Karita she would demand a major input, which was not unreasonable.

Even though John had shown her the places, travelled with her and talked her through some of them he no longer had a role to play. He was redundant! She was launched on her own and now only required the services of admittedly an old rival Tom Nolan to complete her *magnum opus.*

This was a bitter blow. Since the phone call he felt suspended, motionless, in an agony of inertia, a machine without power. He'd not known what boredom was, he was always on the go, always active. Now he felt things coming to a stop. There was something dead inside, which did not want to respond to anything. Nothing to do but bear the stress of his own emptiness.

# 6

Peter Smith arrived home late one evening. A dark and empty house welcomed him and letters lay over the floor. He picked them up; most were bills, circulars and advertisements. Two looked official, one marked Q E 2 Hospital and the other from Addenbrooke's in Cambridge.

He guessed it was about that woman again, Karen, with her constant complaints.

When she first became ill doctors advised a long break, perhaps the west of Scotland, Western Isles or even Skye where the air could be invigorating. Plans were made for the train journey with overnight sleepers. The outward journey was long but relatively uneventful. Peter was very attentive and the night-porters most helpful as they knew she was unwell. They arrived at a recommended boarding house. The holiday improved her health and she enjoyed the air and scenery, the peace and quiet – but he noticed a subtle change in their relationship. At first he paid little attention and put it down to changes in the environment – then it nagged at the back of his mind. Something insidious but inexplicable was happening.

When they were first married she'd a lot of arrears of living to make up. Her days had been as barren as the cream coloured schoolroom where she tried to coerce or even compel knowledge into recalcitrant children. Her trouble started soon after this. Life had a grudge against her and he took pity on her.

Then it was time to return home and all the detailed planning was left to him. She just wasn't interested anymore. He pretended to ignore it. They boarded the train for the first leg of the journey. Each had separate berths in first-class. As they started off he stared at the shadows overhead and was conscious of the rhythmic noise of wheels, which had the effect of driving his brain into deeper circles of wakeful activity -- perhaps it was only returning to the turmoil of newspaper life with its deadlines and chaos. The rest of the sleeping car sank into its peaceful night silence. He peered through wet window panes that showed sudden flashes of light and stretches of hurrying darkness. All was quiet in his wife's berth.

He wondered if she wanted anything and if she could hear him if he spoke. Recently her voice had grown weak and annoyed him when she did not hear. This irritability triggered an imperceptible estrangement. They were like two people looking at each other through a sheet of glass; they were close together, almost touching but they could not feel or hear each other clearly; their communicating powers were gradually being eroded. He had a sense of separation, and imagined he saw it in the look with which she supplemented her failing words. Doubtless the fault was his. The change found him unprepared. Previously, their pulses beat in unison; both had the same confidence in an optimistic future. Now their energies no longer coincided, he still bounded ahead with life full of hope while she lagged behind struggling to keep up. They returned home and life started its humdrum existence.

Weeks later she was admitted to the Q E 2 Hospital in Welwyn for yet another course of treatment. He became sick and tired of her complaints -- they were a millstone around his neck. He'd to get away and it was a relief to admit her to hospital. He had to escape; life was getting intolerable. A conference in Bournemouth materialised and he put his name down as a delegate; yet there were other plans. Hastings seemed more attractive and the company better. So it was arranged and for a short time things went well.

Now he was back home, cold and miserable. No food in the house, the place untidy and unkempt and he felt despondent again.

He phoned his girlfriend. 'Hello Penny. You got home safely then? I said I'd call you.'

'Yes. Pleasant journey,' she replied happily. 'Thanks for the lovely time in Hastings. It was real fun. I'd love to do it again.'

'We'll have to see how things work out.'

'How do you mean?'

'Karen's in hospital.'

There was a gasp at the end of the line. 'What again?'

'Yes, again. And funny I don't know where. There were two letters waiting for me, one from the Q E 2 and another from Addenbrooke's in Cambridge. I'm reluctant to open them. I've a feeling no matter what, they'll have bad news. And that's the last thing I want now. That bloody Karen woman has been bad news all along -- always problems at home and in hospital. It seems never ending.'

'You poor dear, Peter. I wish I were with you. You need company and consoling.'

He replied a little awkwardly. 'I think that would be unwise, Penny. The neighbours are worse than central control night-watch. There are voluntary spies in every neighbourhood. They miss very little. We'd better not take any chances. I think I'll just take it easy and deal with the correspondence in the morning. It's too late to bother anyone now.'

Outside a police car stopped on the road a few houses away from the Smith home. The officer recognised the motor registration and saw the lights in the house. He radioed HQ and reported his findings.

'Hello Chief. I'm parked on Peter Smith's street. I think the husband is home, the car reg tallies and there are lights on in the house. Await further instructions. Over.'

'Well done sergeant. Stay put for a while and do nothing. I'll get word to Cambridge and speak to Chief Inspector Jordan or his deputy. They're in charge. I'll get back to you a.s.a.p. I've any word.'

'Right. Chief. Over and out.'

The sergeant sat quietly in the car making notes.

The radio bleeped again and he answered. 'Welwyn HQ here. I've spoken to CI Jordan who was pleased to hear the news. He advises no further action your end. Don't want to alarm anyone. This is a sensitive issue. Jordan says he'll let Nicholson know we've located Smith. The CI thinks at this stage things should be left to the medics to sort out their own problems and it is not a police matter. You can resume routine patrolling again and forget about the Smith case.'

'OK. Chief. Will do. Over and out.'

Peter Smith had a late breakfast of coffee and stale bread; it was all he was able for. He opened the Q E 2 Hospital letter. It was from Dr. Wilkinson informing him that Karen's condition had deteriorated and she had been transferred to Addenbrooke's Hospital in Cambridge where treatment facilities were better. He hoped Mr. Smith would have no objection and realised the distance for visiting could be a problem. Any further information could be obtained from the consultant in charge, Dr. John Nicholson.

Dr. Nicholson's letter was more direct. He repeated the referral facts and then asked to see Peter Smith as soon as possible. There was an important matter they had to discuss. He put the letter down in disgust. What important matter? Couldn't they just get on with it? He wasn't really needed and the hospital was not just around the corner like the Q E 2 but nearly forty miles away. A day off work! How inconsiderate some people could be.

In spite of himself he decided to phone Nicholson in Addenbrooke's to find out exactly what was going on and why all the fuss. He was going to tell them how busy he was and had little time to spend on long distance travel and probably unnecessary visits to the hospital. He had all this before – repeatedly -- it was getting intolerable.

# 7

Still it had to be done. Peter Smith lifted the phone and dialled Addenbrooke's in Cambridge.

A pleasant female voice came on the line almost immediately.

'Addenbrooke's reception. Can I help you?'

'Yes please. My name is Peter Smith and I'd like to talk to Dr. John Nicholson about my wife who is a patient at your hospital, if that's convenient.'

'One moment please.'

Peter Smith lit a cigarette while he waited.

'This is Gillian Gilchrist, Dr. Nicholson's secretary. Can I be of assistance?'

'Morning, Miss Gilchrist. I don't really think you can help. I'd better explain. I've been away on business for a few days and I understand my dear wife had been transferred from The Q E 2 Hospital to Addenbrooke's without my consent. Dr. Nicholson wrote requesting me to get in touch without delay because there is some urgent business and ... '

'Please hold the line Mr. Smith,' she interrupted. 'I'll see if I can trace Dr. Nicholson.'

'Ah good,' Smith murmured grudgingly. 'Maybe we're getting places.'

He lit another cigarette.

'Mr. Smith are you still there?' It was Gillian again.

'Yes. I'm here, waiting.'

'I've managed to locate Dr. Nicholson and he'd like to have a word. Could you hold?'

Several clicking noises occurred. A soft voice came on line. 'Hello. Is that Mr. Peter Smith?'

'Yes, Smith here. You must be Dr. Nicholson.'

'That's right. I wrote last week, as we could not find you at home or in your office. Dr. Wilkinson in the Q E 2 asked me to take over the care of your wife because the treatment available in the Q E 2 was ineffective.'

'I'm surprised at that.' He sounded irritated. 'Shouldn't it be the same all over the NHS?'

'The simple answer to that question is it should be but it isn't. There are many examples that could be quoted. Politicians would deny they ever happen but they do.'

'I see. You've made some interesting comments there.'

'However, in your wife's case things are somewhat different. It's difficult to explain over the phone and perhaps I should not even attempt it. In her case the situation in Addenbrooke's is better than elsewhere because of our research mechanisms.'

'Are these not available elsewhere such as the Q E 2?' He asked indignantly.

'No they're not. And there are good reasons for this.'

'I don't understand.'

'Mr. Smith. I think the best thing is for you and I to have a face-to-face discussion. The sooner you can visit here the better. I'd appreciate an early visit.'

'Dr. Nicholson, do you realise I'm a very busy man.' He was getting irritated now. 'I've a load of deadlines to meet, my days are full and I've little time to be traipsing off forty miles up country to yet another hospital just to be told that my wife is getting the best treatment but she is having difficulties. I've heard all this before so often it makes me sick.'

John drew a deep breath. 'I'm sorry to hear you feel that way about your wife. She has been making good progress on her new treatment and I thought you, at least, would like to see her.'

There was a silence that puzzled John.

'To see her, you say. How will she be? The same human cripple capable of doing nothing. I'm not sure I want to know.'

Like most well-mannered individuals, John Nicholson had no protection against the rudeness of others. Although these comments did hurt, he lacked the brusqueness to give a direct snub. He merely replied.

'Look here Mr. Smith, whatever differences you may have with your wife they do not concern us. These have to be sorted out between yourselves. You must remember we've a job to do and we try to do it to the best of our ability. That I assure you we are doing.'

He paused for a moment waiting for a response. None came.

'Your wife has done reasonably well on the new treatment. She's in less pain and has less breathing difficulties. She even mentioned she would welcome a visit from you if you could possibly find time. How about it, Mr. Smith?'

A sigh full of frustration came over the line.

'OK. I'm occupied this weekend. I could make it about two o'clock on Monday next.'

'That'd suit me fine. When you arrive please ask for Gillian my secretary. She'll direct you to my office. I'd like to speak to you first before you see your wife.'

'Dr. Nicholson it all sounds very intriguing. I get the feeling there's more going on than meets the eye. I just wonder what it is. I do hope you will be open and frank with me. I don't like a cover-up. As far as I'm concerned it'll always be found out and there could be trouble.'

'Mr. Smith, there're no cover-ups and such like. I'll explain everything to you as clearly as I can and you can be your own judge. I think we'd better leave it at that. See you on Monday.'

'Right. On Monday then.'

What a strange character this Mr. Smith was. No love lost for his wife and full of questions about treatment and ethics. John had a feeling of problems ahead.

*

Meanwhile in a little village called Port Isaac in Cornwall Graham Sinclair and Helen Watson relaxed in the balmy sunshine as they walked the many cliffs and sea paths. They were happy to be together and decided, like Robert Frost -- 'I took the one (road) less travelled by, and that has made all the difference.'

# 8

Karita Isselherg waited patiently for Tom Nolan in the lobby of the Shelbourne in Dublin. A meeting was arranged for 11.30 a.m. followed by a working lunch. She phoned him the previous week with news about the proposals. Tom's wife, Ruth, was guarded yet reluctant to speak her mind. Nevertheless Tom agreed to meet Karita to get the project rolling.

He was right on time. Karita reserved a table in the lounge. It was loaded with books, maps, and reference leaflets. It would not be appropriate to invite him to her room although it would have been more convenient and relaxed. Not in Dublin, though. She'd have done it at home but here she respected local conventions.

'Thanks for coming, Tom. Now we can get down to business. I tried to explain the outline of the proposal over the phone but it's much easier if we have all the references in front of us. We can plan an accurate itinerary.'

'Sounds logical to me Karita.'

'The first thing you should have is a copy of the manuscript. It'll give an idea of the atmosphere of olden times and you can steep yourself in the goings-on of the period. How should we begin?'

Tom rubbed his chin and looked across the room.

'I could begin with a number of questions,' he replied thoughtfully.

'Such as?'

'Let's see then. What, how, where, when and why do the illustrations?'

She laughed. 'OK. That will do for a start. The list should include the Sligo countryside, the area around Ballycastle and the Ceide Fields, the Dun Angus Fort in Aranmore, Doo Lough and Mweelrea Mountain, Croagh Patrick, Staigue Fort in Kerry, the Gap of Dunloe and Ballaghisheen Pass along with the Skelligs, the rocks at Mizen in Cork and Tara Hill in Meath.'

'What a list!' He was flabbergasted. This was just a start.

'The how I'll leave to you,' she continued. 'But I'd like to suggest we go on a tour so that I can have some input into the views leading to the compositions. I mean no disrespect but I can make suggestions and you can modify or reject them. At least you'll know how I feel.'

'I'd welcome all suggestions. When do we ...?'

'As soon as possible, Tom. I'm working to a deadline and the sooner the better.'

'Fine. I can be ready in a few days. I'm a free agent and this commission is important.'

'I'm glad you see it that way. Yes. It'll mean publicity for you. All your expenses will be paid and each of the finished works will probably be purchased at gallery prices and held in the library of Kungsangan House -- to begin with -- where most of the writing was done.'

'The last question is why?'

'That's the most difficult one to answer, Tom. I'm convinced no photograph could capture the life and times of these places like you can. It's a matter of imagination; mine has gone into the writing and I'm confident you've plenty of your own.'

Tom smiled.

'Let's look at the maps.'

Before Tom left the hotel she handed him a cheque for £7,000.00 to cover expenses.

# 9

Ruth Nolan went up stairs to her bedroom and locked the door. She wished to be alone. Looking in a long mirror in the corner she was still slender and quick on her feet, although in recent times there had been no need to rush as much as she used to. Domestic duties had become routine and were carried out at a comfortable pace. Yet, she felt compelling impulses of youth revived as a result of the news.

She sat in front of the three-faced mirror on the dressing table – a treasured present from her mother. The features reflected back did not seem suitable for general display with signs of emotional upset, blushes, blemishes and untidy hair.

Her fair hair was sparse, traces of grey appeared at the temples and the lips thin and unyielding. A twitch suddenly became noticeable, which she was previously unaware of, and at the corner of the eyes cracks were apparent. This was a face that had imperceptibly grown middle-aged while it yearned for the pleasures of youth.

Yet she could still blush. She drew back to get a less detailed appreciation of her appearance and the throbbing blush threw a pleasant veil over the paleness of skin, the light lips and the incipient cracks at the eyes. How a little colour could help, which she applied delicately – not too much -- here and there. The eyes became deeper and luminous and more mysterious.

The edge of her dress was also a little low; it showed undesirable lines in the neck. She looked around for some camouflage and found a length of rose coloured velvet and wrapped it around her neck. Yes, that was much better.

Behind all this was the news about Tom, and Karita Isselherg. She was upset when she heard of his plans to go on location. What was the point? Could he simply not get a list and visit the places alone and make his own mind up – as he always did? For years it had been his modus operandi and it worked well. Why should it change now?

Tom had listened patiently and felt sorry for her. He'd explained this was only a business deal – strictly business – he'd report regularly on progress and it would mean a lot to his career if the illustrations were accepted with world-wide distribution.

The final evidence that convinced her was when he produced a cheque for £7,000 as an advance for expenses and there was an option of buying the completed canvases at gallery prices for a special collection in Sweden. She was reassured and apologised for causing a fuss; it was because she loved him and wanted nothing to come between them. She could trust him completely.

Tom had a few days to read Karita's incomplete manuscript. He found it absorbing, complicated and fascinating. He'd never read an approach to Celtic mythology like it. It was alive with characters and her descriptive powers were compelling. She certainly knew her subject. She phoned three days later saying she'd made the arrangements and the car was already hired. Was he prepared to leave tomorrow? He was. Then the first port of call was Sligo.

He gave a little nervous laugh. 'That's up to you, Karita, you're the organiser and you know what you want.'

Next day they set out towards Mullingar. As they travelled she talked about the project.

'Tom you must realise that prehistoric myths about Ireland

are endlessly complicated. One cannot take them too seriously; otherwise despair grips the literal mind. Yet some history must be present. Scholars must sift history and myths together and hope some truth will come out. A lot is present in Sligo – both mythology and architecture -- evidence of Minoans, the Fir Bolg, Formonians, the Tuatha de Danann and Fianna. Every trip from Sligo town can take us back several thousand years almost to the birth of Irish civilization.'

He made a helpless gesture. 'Karita, that's heavy stuff and leaves me a little confused. However the weather is good so if we travel around I'll know a gem location when I see it. As for mountain climbing – no way! It's from the fields and valleys I see great things. They're only small and tiny from the top and anyway I'm usually tired and cold to do good work.'

After visiting the Ceide Fields Tom said.

'This is amazing, Karita, but I don't think I could ever paint such a scene.'

'That's right Tom. This is the purpose of the whole exercise. When you see the place and what's in the manuscript then you can take or leave it. You must only select what appeals to you. There's no other way. Otherwise your work would suffer. You are not subject to rules and regulations. You're a free agent.'

'Thank God for that because I'm confused so far. Your manuscript alone is so complicated and I mean that as a compliment. I'm a simple-minded man that sees things quickly or not at all.'

Stague Fort near Waterville was visited and she asked him to read about it in the manuscript as it was a significant part of the storyline. Ballagisheen Pass held a special place in Karita's emotions and she hoped Tom would be able to catch the awesome beauty of the place looking eastwards towards Carrantuohil – her favourite location.

The final part of the trip was Mizen Head. She asked him to walk across the foot bridge to the light house, look north at the cliffs and see if he could do a painting of what he saw. He agreed. She remained behind and waited for him to return – safely.

# 10

In the hotel that night Karita made notes for The Trilogy. They were rough and would need fine-tuning but, at least, she'd the energy and willingness to complete the task. She wrote quickly as various aspects seemed to gel.

... A consideration of the making of myths and legends could start almost anywhere and any place so why not begin with some generalities? No apology will be made for choosing a geographic basis at random. It has been claimed Irish history started in Ulster. The earliest evidence of human presence comes from around 6,500 BC in Colraine near Derry – a short distance from Scotland – and ever since the country has made more history than it could ever consume ...

... In ancient times the land was run by three powerful Goddesses who established a glorious mythology, unsurpassed in extravagance and beauty by any other European country except perhaps Greece. A beautiful literature, an unparalleled design in jewellery and ornament in stone, bronze and gold, only a fraction of which survives. They were recognised for extraordinary schools of thought and magical religion, whose effects can even touch our lives today...

... A lot of these points have been detailed in the main body of the text and so it only remains to clarify recent history and mythology without threading on too many toes ...

... In more recent years few countries have produced little work of outstanding merit on her own soil. This was achieved in other lands across the globe. Why? Perhaps some blame lies in the obstructive nature of those who placed themselves in positions of power to say 'no'; somebodies who knew better than allow iconoclastic works to be let loose on a pure and untainted people. Jealousy, perhaps, may have been a motive but not mentioned let alone thought of. A far more decent word is censorship. Such a clean, clear and well-meaning word that could satisfy any committee meeting and gathering of clergy. Religion and history could always be mixed together and there was no arguing with them. It wasn't the committee's fault but a much higher authority delivered by sacred hands and script ...

... One major contribution was geography. No other European island existed so deplorably close to such a hostile enemy, who had for centuries no neighbours on its east but her conqueror, and nothing out west but a four-thousand mile stretch of ocean to America. The isolation and forbearance of its people must have known no bounds ...

... And yet some say that Ireland is amongst the most beautiful countries in the world. What would Monet or Van Gogh have made of the wooded cliffs that appear to be carved by a magic hand out of precious stones, multicoloured mosses stained with rare alpine flowers, dragon flies, golden fairy rings, deep silent and deserted woods honeycombed with sunshine? ...

... Yeats describing Ireland's development felt that its most inspiring moment lingers at twilight – that special time between day and night, light and dark, resting and sleeping – time so precious to all it can disappear more quickly than we think; it is at this time we can really feel the pulse of Ireland if we listen for it. No one can really escape it – ignore if we will! We cannot blow it away as it descends on the collective Irish mind usually when work is done for the day. And the magic is ours if we let it. Perhaps children in innocence realise it's a special time and wish it would never end ....

... Other Irish people, whether a success or failure, will be sitting on a bar stool putting a false glow on things, or be down on their knees praying with wordless thoughts or even looking out windows as the world gets darker. Maybe they can see the towers of New York or Liverpool or London and imagine they are the mountains, rivers and lakes of their youth ...

... Attempts were made in this Trilogy to awaken in readers, and especially in those who know something of history, dreams of a deeper wonder that seem like an earthly veil. There is an abundance of ruins, great and not so great scattered throughout the land and evidence of an amazing art and science of a highly educated population has been found. Then there are skeletons of cottages lying on the side of mountains and in the most inhospitable places where living – if that's not a misused euphemism – must have been a struggle. It wasn't only the weather that gnawed and clawed away at these miserable dwellings that families called home. Alongside these there's evidence of lazy beds. Surely, there has never been a description of structures that carry a more wretched and meaningless irony. When they and other sources of food gradually disappeared the awesome figure of Famine came on many far reaching visits producing a fine crop of its own. Many millions disappeared from the land either locally or scattered to God only knows where ...

... Different stories are told about those who survived and prospered. For this reason recent legends and history can be a minefield. Nobody is completely right or completely wrong unless it is said so by the opposite side. There's the dilemma ...

... The nearer one gets to the end game the more difficult the task becomes, not because facts are not clear – they are clearer; not because they are harder to find – they are more evident; not because of more uncertainty. No! Truth shines through more clearly in spite of attempts at obscuration, deceit and cover-up for whatever reason ...

... I have found Ireland a country of the strangest contradictions and because of this she is not unique, although it would appear that nowhere else in Europe are there such extremes of waste and brilliance, poverty and riches, apathy and vigour, foolishness and intelligence, callousness and caring, mental blockage and cerebral clarity ...

... The weather may have something to do with it – as they say – or it may indicate a forgotten past and, perhaps, a forgotten wisdom ...

Karita stood up from her desk and gently blew out the candles.

# 11

Peter Smith had difficulty in finding Addenbrooke's Hospital. He was not a frequent visitor to Cambridge and his mood not the best. Eventually, when he found the car-park most places were full fuelling his agitation. Fifteen minutes later he was successful.

He grabbed a briefcase, banged the door and hurried up the steps to Reception.

'I believe Dr. Nicholson is expecting me. My name is Peter Smith and I've travelled from Welwyn Garden City.'

'Just a minute, sir. I'll phone his secretary and she'll look after you. If you'd like to take a seat over there. She won't be long.'

He looked at his watch, made his way to the seating area where The Guardian Newspaper lay on a coffee table and a faint smile came to his face. He hadn't long to wait. Gillian's voice came from behind.

'Mr. Peter Smith, I assume. Dr. Nicholson would be pleased to see you now, if you'd like to come this way.'

'Fine. You lead the way.'

They walked several corridors and arrived at a door with a brass plate with Dr. Nicholson's name on it.

Gillian knocked and John opened the door.

'Ah, Mr. Smith. Nice to see you. Glad you were able to make the journey. Hope it wasn't too arduous.'

'Arduous enough. It's a long way from Welwyn where it was easier to visit my wife.'

'Please make yourself comfortable. Gillian will bring tea or coffee.'

'Coffee please. White. One sugar,' he said grimly.

'Very good, Mr. Smith. It will be right along.'

She left the office.

John sat in a chair beside him. 'Now Mr. Smith. Down to business. I've asked to see you to discuss your wife's treatment.'

'Doctor. I'm still confused. Why on earth can't she be treated in the Q E 2 Hospital?' Smith asked intervening. 'She's been in and out of that hospital for years and seemed to be well controlled. I can't understand how or why she ended up in a place like this.'

Suddenly John was serious.

'I don't know how much Dr. Wilkinson told you but several weeks ago Karen had a set back, a significant one and Dr. Wilkinson, who's aware of our work here asked for help. At the time there was only one option and that was to transfer her to Addenbrooke's for full assessment and to see if our treatment could halt the deterioration.'

'I see.' Smith replied grudgingly.

'This protocol is only available in Addenbrooke's,' John said evading his eyes. 'We have pioneered a technique that's still in the experimental stages.'

'Even so why is it not available elsewhere? I've asked this question before but didn't get a satisfactory answer.'

'For a long time Cambridge has played a leading role in science, Mr. Smith, as you're probably aware. I can see you are a well educated man.'

'Thank you, Dr. Nicholson.' Smith answered with irritation. 'But flattery won't overcome my reluctance to accept why my wife cannot be treated locally. It's a great inconvenience to get here and I'm a very busy man.'

'I should explain the full facts,' John was patient. 'Your wife was reasonably well controlled on conventional medication, which is produced by drug companies for the mass market and not the individual patient, although certain dose adjustments can be made. These are limited. What we're doing is removing the patient's cancer cells, together with normal cells, growing both in tissue lines in the laboratory and almost tailor-making exactly the right drugs to target the cancer cells and have the least effect on normal cells.'

'Sounds logical. What's it called?'

'The in vitro technique.'

'Why?'

'Because the resultant drug, which is a modified chemical agent, we decide to use has already proved itself in the laboratory in cell lines and doesn't have to be tested on animals yet again. We regard this process as long winded and old-fashioned.'

Peter Smith raised a hand. 'Wait a minute. I thought all drugs have to be tested on animals before they can be used in humans. Is that not so, Dr. Nicholson?'

'Convention dictates that they should -- and I use the word advisedly -- but we believe we've gone a stage further and bypassed animal testing for a second time.'

'Don't you have to license your drugs stating they have been tested properly?'

'To a limited extent that's true and they already have been initially. Also we have overcome the problem in Addenbrooke's by getting the patient's consent to use our protocol and legally this seems to work.'

'You mean written approval, Dr. Nicholson?'

'Yes. That's right.'

'Did my wife give written approval for this treatment?' Smith was implacable.

John Nicholson hesitated. 'No. She said we would need your approval. She seemed frightened that anything she signed would not be approved by you.'

Smith banged his fist on the table. 'She's damned right. I don't like the sound of this at all. Experimental drugs being used on my wife without written approval and these drugs are unlicensed. Dr.

Nicholson this is shocking news to me. I can hardly believe it. How could you have gone ahead without approval?'

'We had her verbal approval.'

'Knowing her she was brainwashed. I've always told her to get my permission first to anything as serious as this. I think it is a disgrace.'

John gave a shrug. 'You've got the wrong picture, Mr. Smith. I hope to convince you your wife is getting the best treatment possible and she's making steady progress since her admission.'

'Sorry for my outbreak doctor,' Smith said, calming down. 'I've been under a great strain recently, many worries and now all this trouble and upset with Karen. She can be a handful, you know.'

'How long have you been married, Mr. Smith?'

'About seven years. And she's been sick on and off most of that time.'

'I can appreciate how you must feel. The strain of chronic illness can be intolerable. May I suggest we go along and see her now?'

'That'd be a good idea.' Smith sounded a little reluctant.

'We have her in a single room at present. It's a bit lonely at times but it helps a great deal in barrier nursing. Infection is becoming an increasing problem. So she's in the best place.'

'Glad to hear it. Please lead the way.'

John decided to postpone asking Smith to sign the consent forms for the present. He would approach him after he'd seen his wife and before leaving the hospital. It was a five minute walk to the Clinical Oncology Unit and Peter Smith was a little confused about directions.

'Don't worry. The Unit is close to the main reception area. However, I'd like to see you again before you leave the hospital. Here we are now. We'll go straight in.'

John gave a gentle knock on the door and slowly opened it. Karen was lying in bed, with eyes open and staring at the ceiling.

'Hello Karen,' John said gently. 'We've a surprise for you today. Peter has travelled from Welwyn to see you. Isn't that nice?'

She turned her head and looked at her husband whom she hadn't seen for weeks. He bent down to kiss her on the cheek but she turned away. No smile. No welcome. It was obvious she wasn't pleased to see him.

'Well Karen. How are you getting on? I've been hearing good reports from Dr. Nicholson here that you're on the mend again.'

There was silence as he sat on the chair beside the bed and tried to hold her hand. It was withdrawn slowly and she stared at John Nicholson with a frightened look, appealing for something and then closed them.

'Look. I think I should leave you two together,' John said a little hastily. 'There are probably things you both want to catch up on and I'd only be in the way.'

'Fine,' Peter Smith said. 'And thank you for the coffee and explanations. I found them intriguing and very interesting. I'd like to follow them up in greater detail at a later date, if you don't mind doctor.'

'Always glad to help.' John turned to Karen. 'I'll probably call in later this evening, if not early tomorrow morning. Goodbye for the present.'

'Dr. Nicholson,' Karen replied, 'do you have to go so soon?'

'Yes Karen. I'm afraid I must. And it's best that I do so. Don't worry I'll see you later.'

He shock hands with Peter Smith. 'Remember please come and see me before you leave the hospital.' He then left the room.

Sister Mortimer spoke in a hushed voice into the phone. 'Dr. Nicholson. I must talk to you urgently. Mr. Smith stayed with his wife for about an hour and a half and I made several excuses to visit the room to see what was going on. Apparently they were having a frightful row about something -- not much to start with -- but on my third visit Mr. Smith was in a highly aggressive mood, I could see poor Mrs. Smith was in a distressed state. She was hysterical and crying. I had to ask Mr. Smith that the visit should end because of the effects on the patient. He agreed to leave, abruptly banging the door and saying no words of goodbye. I checked her blood pressure and it was dangerously high so I think you should try to calm her down. The poor dear is so distressed over something but she won't tell me what.'

'All right sister. I'll come over about seven o'clock after this paper work and before you change over to night staff. What a predicament. Exactly what she doesn't need just now.'

The University Grants Committee report was causing lots of trouble for John

Nicholson. There were so many interruptions and interferences he'd have to take the papers home to get a bit of peace and quiet. And in the back of his mind there was the Karita and Tom Nolan proposal. He really didn't know why. And Peter Smith had disappeared before signing the consent forms for his wife.

# 12

**P**eter Smith had time to reflect on his return to Welwyn. As an investigative journalist he thought he could smell a story in the goings on between the Q E 2 and Addenbrooke's. If more information was forthcoming he'd have to be as nice as possible to everyone who could be useful including Wilkinson and Nicholson. There could be a big story there -- perhaps an exclusive. Everyone's dream! He'd play his cards close to his chest as he was a seasoned pro. Just leave it to Pete Smith, they'd say; he'll get to the bottom of it.

About seven o'clock John Nicholson walked into Sister Mortimer's office.

'How glad I am to see you, Dr. Nicholson. Please sit down,' she said breathlessly.

'How's Mrs. Smith?'

'No change in spite of the antihypertensives you prescribed three hours ago.'

'I'm sorry to hear that,' he said with a sigh.

'Please forgive me Dr. Nicholson, but that brute of a husband seemed to have a dreadful effect on her. They were fighting all the time. At least he was fighting with her and she was just lying there taking it as best she could.'

'Any idea what it was about?'

'I couldn't help hearing references to him being away from home and nobody could find him. Where was he? No answers. Then he attacked her for giving consent to things she could never know about in a hundred years. All these were serious things and she replied that he was nowhere to be found. She insisted on getting answers to her questions. All he would say they were secret assignments with his work and could not discuss them. She would not believe him and so on. Each shouting. Each trying to win points. In the end I had to come to Mrs. Smith's rescue. She was in a state of collapse, which he didn't seem to notice or care about. I insisted he leave because she was so distressed.'

John smiled reassuringly and said. 'You did well sister and the right thing. Perhaps we should go and see her now.'

'Of course. She's a little sedated but still on edge. I've never seen her like this since her admission.'

'Poor thing. I do feel for her. And I wonder what's on the husband's agenda.'

'Mrs. Smith. Dr. Nicholson is here to see you again before he goes off duty.'

'Hello, Dr. Nicholson. I'm sorry there was such a fuss this afternoon with Peter. He's inclined to get excited about things and it's hard for him to see reason. I warned you about him.'

'Indeed you did. I had a taste of his temperament before we came to see you this afternoon. He seems to question everything -- everything we are trying to do. I was as helpful as possible although it did get irritating at times.'

'Doctor, you put it so mildly. With me he becomes completely impossible if he does not get his own way -- all the time. For example, he was raging with me for coming here from Welwyn because it was inconvenient for him and then he was more incensed that I agreed verbally to your treatment even though he didn't know the full facts. Most unreasonable. He told me I knew nothing. Never did. And before he left he said he was going to get his own back on me. That last remark has me so frightened. He's unpredictable. He's capable of doing anything, if anyone crosses him. Please help me Dr. Nicholson. I'm afraid of him and perhaps something else. I just don't know what. And now I've got this terrible headache. Although I'm not surprised after all that. I never get headaches.'

John turned to Sister Mortimer. 'Please check her blood pressure again.'

This was done. 220 / 180. No evidence of reduction.

'Mrs. Smith.' John said gently. 'Your blood pressure is still high. I think we should transfer you to ICU for a day or two. Please sister could you arrange this straight away.'

Some catastrophe appeared to be looming on the horizon that he had no control over. He would just have to wait and see what developed and cope the best he could. No matter what.

# 13

John Nicholson's senior registrar, Dr. Egan, telephoned Peter Smith in his home in Welwyn. No reply. He tried The Guardian Newspaper in London. Eventually he got through. He said gently.

'Mr. Peter Smith. Addenbrooke's Hospital here. This is Dr. Egan, senior registrar to Dr. Nicholson whom I believe you've met. He asked me to tell you you're wife has deteriorated since your visit last Monday. She had several episodes of hypertension, or high blood pressure, and difficulty in breathing. It was hard to control in the wards so she was moved to ICU to stabilise her. She's still not well.'

Peter Smith paused for a moment. A little restraint was called for.

'Thank you, Dr. Egan. It's kind of you to keep me informed of Karen's progress. I don't know how I can get out to see her again. It's a long drive from London. Anyway, I know she's in good hands.'

Then a thought came to him. 'One question, Dr.Egan. Is she receiving the treatment developed by Dr. Nicholson?'

'Yes, she is,' he said simply. 'It seems to be the only way to keep her reasonably stabilised.'

'I see. Interesting. I must remember that.' A smile appeared on his face. 'Now will you excuse me? There are lots of problems facing me and time is precious. I do appreciate the call.'

'Not at all. My pleasure.' Egan said in a courteous voice. 'Goodbye Mr. Smith.'

The scene in ICU was frightening for Karen Smith. She'd heard about it but never experienced the drama and activity that went into running such places. Each patient was linked to six or seven monitors at a highly complicated looking bed – which was more like an operating table. A central control panel was fed from the equipment and flashing lights and gentle sounds emanating everywhere. However there was a general feeling of calm and with almost one nurse per patient.

As she lay there she felt confident and reassured. Here was the nerve centre of the hospital where all stops were pulled out to save life; no rushing through traffic, no roadside drips and no wrangling with people getting in the way. She'd known so many aspects of hospital life over the last seven years. Here she'd arrived at the epicentre of something – something important. She was important – almost for the first time in her life. Every need catered for. She didn't have to ask for anything because a nurse already anticipated her requirement. It was almost like a little piece of heaven. No pain, no worries and no stress. She almost wished she could be here forever with lots of ministering angels around -- a sublime experience being wrapped in cotton wool.

She spent two days and nights in this haven of luxury. At the end of the third day she began to feel unwell. A frontal headache gradually developed but she didn't pay much attention. Headaches she had before so why report them? Although this was different. It was just behind her forehead and, at times, it caused interruption in vision.

The duty nurse noticed she'd dropped off sleeping and left her for fifteen minutes. Then it was time to check her blood pressure according to the schedule. To her surprise it had dropped so she alerted sister in charge. Karen was not in a natural sleep but semicomatose. The pupils were dilated and slowly responding to light. The senior registrar was called immediately.

Dr. Egan completed the examination and after speaking to Dr. Nicholson they decided it was a neurosurgical emergency. She was undergoing a subarachnoid haemorrhage and there was only a slim chance surgeons could operate in time to save her life.

An attempt was made to clamp the bleeding vessels but the fulminating nature of the episode was too much. Karen Smith died on the third day of admission to ICU in Addenbrooke's Hospital.

The husband had to be informed as soon as possible.

# 14

John Nicholson sat in his office opposite Peter Smith discussing Karen's death. It was difficult for both. Smith's face was flushed and the bright eyes avoided John's most of the time.

The sensual lips were wide and expansive and, although, silent at present John guessed he'd a lot to say. It was his profession – interviewing people, consistently attempting to extract maximum information -- sometimes for devious reasons. He would have to be careful.

John was puzzled about the other's intentions. Was it to be a straightforward account of Karen's last days? Or was there something else on his mind? He could only see a stranger, inexplicable, with an ill-defined pattern of passions, desires, enthusiasms and above all an inquisitiveness that was unsettling. If Smith could ask three questions on any subject instead of one he would blast ahead. A feeling of endless probing was infused into this slender man of nearly forty years who probably was the same as he was at twenty. Still he was a loose cannon and a devious personality -- all probably a reflection of his profession and even the way he lived.

'Dr. Nicholson, may I ask exactly how Karen died?' Smith inquired in his most pained expression.

'The immediate cause of death was an aneurysm in the Circle

of Willis in the lower anterior part of the brain, which ruptured and produced a subarachnoid haemorrhage.'

'A brain haemorrhage you mean?' He asked dissatisfied.

'Yes.' John answered flatly. 'It was so sudden the neurosurgery team couldn't gain access quickly enough to clamp the bleeding vessels.'

'I see. And what could have caused the aneurysm?'

'Mr. Smith. It's a medical fact that blood vessels in this area in women are more prone to haemorrhage than in men. It has been known as "the weaker link" in women, perhaps the only one because women are generally regarded as healthier than men in the totality of things -- although matters have been changing.'

'And can you think of any circumstances preceding the brain haemorrhage that could have lead to the bleed?' Smith persisted.

'She developed high blood pressure, which was difficult to control. This only happened in the last week and commenced after your first visit to see her.'

'Are you implying I had anything to do with it?' He demanded.

John looked straight at him. 'No, not implying. Merely making a statement of fact.'

'Could it have something to do with the treatment she's received since been transferred to this hospital?'

'That's unlikely.' John was losing patience. 'We've not noticed it in others receiving the same treatment and of course we watch out for it. It's more likely she was emotionally upset on that Monday. Could that be the case, Mr. Smith?'

'We had our differences. Who doesn't? But it was nothing. She blew things out of all proportion and, of course, she was very ill at the time. I hadn't really seen any improvement in her physical condition compared to other times. I visited her regularly in the Q E 2 near our home.' He gave a significant sigh. 'I still can't see why she'd to be moved to here. She was doing so well there as far as I could judge. I'm sorry to repeat that, Dr. Nicholson. I was her husband and I could see things, that, perhaps, others couldn't. I find the whole business terribly distressing.'

'I sympathise with you – especially now. If there is anything further we can do please do not hesitate to ask.'

'That's kind of you, doctor. I've now to make arrangements for the funeral – probably back in Welwyn. And then carry on as normal.'

'Yes. These tragedies come to all of us sooner or later.'

'And you see more than most, doctor. I don't envy you your job for the world.'

'You know, Mr. Smith, I wouldn't change it for the world.'

With that Peter Smith took his leave. There were a lot of things to be done before the story went cold.

# 15

At the funeral Peter Smith sat in the front pew with his head forward, yet he was aware of an almost full church. After the service he walked down the aisle behind the slowly moving coffin, head still bent yet conscious of faces extended towards him, trying to get him to acknowledge their presence. Suddenly he recognised Penny – Penelope Harvey!

This was appalling – a terrible shock. She should not have come – but stayed away. It wasn't right especially under the circumstances. And Welwyn was a relatively small town – although called a city. People tended to know a lot about one another. More than you think. How dare she? The talk, the gossip, the humiliation.

Karen had known about her for some time and there were many rows. He now saw Penny as Karen had seen. And he despised her. It wasn't decent her coming to the funeral. Maybe she thought she was helping, but this was the height of stupidity and he'd never forgive her.

Throughout the day his fury grew. It filled the empty time after the funeral, the hotel reception, the farewell to guests and well-wishers. The hotel reception was boring and humdrum. It was exceptionally noisy and waiters were generous with their offerings. He was abstemious so a moment arrived when he couldn't take it any more. Several attempts at departure were hindered by nuisances. He was ambushed by a tiresome bore who way-laid him with tears and cornered by a woman burdened with all the woes in the world.

Some friends offered to drive him home. He accepted but when they suggested keeping him company for a while he refused. He wished to be alone in the solitary home and gather his thoughts.

The place was now emptier than ever. He lit a cigarette, retired to the sitting room and his favourite armchair. For ages he sat in a daze with thoughts about the past – but Penny was a persistent intrusion. The last thing he needed on this grimmest of days. What would people say? What were they saying at this moment? His imagination ran riot and his indignation knew no bounds. What was he to do to repair any damage, or what could he do but let things take their course?

He must take action to settle his account with her. There were always the letters from her he kept hidden from Karen. Now was the time to send them back. Perhaps without a word. Let action speak for itself. Even though his profession was one of words, dependent on them to influence and control, he felt that silence could send a more powerful message than mere words. If the letters were to arrive back with no communication from him she'd surely draw the right conclusion. Penny had a passion for going over things repeatedly and exhaustively that sometimes he'd to leave her to cool down. He was not prepared for a torrent of words and a Niagara of tears at the present time. He'd had enough talking and his well of questions had run dry. In this respect Penny was as bad as poor Karen with their endless rows. Even in the hospital in Cambridge. Karen would not give up until she had the last word and, yes, poor thing did have the last word.

He visualised the scene when the letters were returned with no explanation; tiredness disappeared and his pulse quickened. They must be returned without delay but the task could not be entrusted to just anyone. It would be done by himself and the sooner the better. He bundled them up in a big brown envelop. The reason there were so many was, of course, the difficulty in meeting. There were other ways and means and the recent week in Hastings was delightful but followed so quickly by the tragedy in Addenbrooke's. Fate worked in strange ways.

However, a new thought struck him that in returning the letters without a word might give her an opportunity to misinterpret his real intention and give her an opportunity to take up where they left off. Certainly not what he wanted. He found a sheet of headed notepaper and scribbled a message – blunt and to the point – and inserted it into the envelope.

Her apartment was only a mile and a half away so he decided to deliver it by hand. He welcomed getting out of the house for a walk as the air was brisk and dry. It was a moonless night and few people were about so delivering the package was the safest of all. When he arrived there were lights on in two rooms.

He rang the doorbell and Penny answered. She was alone and invited him in. At first he refused and said he merely wished to return some books. But she was insistent. The cold night had sent people home early and there was no one in sight. She persuaded him to come in for a short while and, against his better judgement, he agreed. He'd so many things to do and little time to do them but her presence was hard to resist as always. He was torn.

The thought of Karen rushed into his mind. When she found out that Penny was also in the background and she was losing Peter's affection, life must have seemed hopeless. Karen was probably like him, secretly suffering from the same hesitations and inhibitions and found it difficult to express affection. She would have found it a burden to carry a jealousy that seemed to have no resolution and had no way of overcoming the internal violence it would have provoked in her. Yes, they had rows, but they led nowhere. The only solution was to give in and give up. Which she did. She was now out of the strife, the worry and anxiety and even the shame.

These thoughts came back as he followed Penny along the corridor and into her small, cosy sitting room. There was a gas fire glowing with a small colour television set beside it. The lighting was soft and seductive and a large bowel of flowers sat triumphantly on a table. Even entering the room was a violation of Karen's memory;

but mixed feelings returned with intolerable acuity as Penny's slim figure preceded him, inviting him to sit down.

True passion ought to be open and free, reckless and audacious, uninhibited by any wife's feelings or jealousies, free of newspaper regulations, friendship of colleagues, the horrendous fear of losing one's job or livelihood or destroying one's career. But nowadays, these considerations were uppermost in a lot of 'right thinking' people; it was only the super-rich or super-careless couldn't give a damn about their own or, for that matter, anyone else's behaviour. Having a conscience was indeed a problem and occasionally a curse for the true enjoyment of life – a good life before it was too late. Now perhaps it was too late and left him confused and bitter and wanting to hit back at something. No matter what that might be.

It dawned on him, all too clearly, that the love he'd given Penny was almost as grudging as he'd given Karen.

He would have to try and get away. Perhaps he could get an overseas post to some hot climate where one could lie all night listening to waves chattering with palms and warm trade winds coming from nowhere. One of the festive islands where love and marriage were not regarded too seriously and one could come and go as one pleases, where one could follow one's natural instincts without calling down a personal tragedy or suffer severe degrees of personal degradation. Perhaps there were places where women did not argue, or worry or reproach one with guilt and hate and dramatised everything, where things ought to be as simple as eating and drinking. Surely there were places like that. There had to be.

As they sat down she said simply.

'You wanted to talk to me?' Her voice was gentle and soothing.

'Did you expect me to come?' He asked timidly.

'I don't know. I really didn't know what to think.' She replied and he caught a faint smile in the soft light. 'I suppose I wasn't expecting to see you for a while at least – until things cooled down and some normality returned -- if that could ever happen.'

The comfort of her presence was reassuring.

'I didn't expect to see you. I just felt like taking a walk,' he lied.

'Poor Peter. You poor dear. You needed consoling and a listening ear and how I've missed you.'

Yes, she was trying her best to be helpful, to sympathise. Anything else would have been unbearable. With a glow of satisfaction he appreciated her efforts, whether they were genuine or not. Her empathy made him brave and courageous. She knew how to handle him, or perhaps he'd played down his own sufferings when he thought of hers.

'These past few months have been ghastly, Penny,' he admitted with a sigh.

'Don't I know. You poor dear.'

She sat back on the sofa and gazed into more space than the room itself.

'Peter. I'm so pleased you had the courage to come and see me,' she said in the same restoring tone, 'because I'm leaving tomorrow.'

'You're leaving town tomorrow?' He was startled and overcome with a kind of weakness. Where could she be going under these circumstances? If he hadn't decided to visit he would have known nothing of her intended departure.

'Peter, you must appreciate it is best this way. I'm going to stay with cousins in Cornwall. They have always been asking me to stay for holidays or longer and I get on so well with them. They are kind people and I'm very fond of them. Mother and father think it would do me the world of good and it'd be a wonderful change from the way of life up here. The climate is mild all year round.'

He could think of nothing to say. Suddenly life looked empty and void.

'I didn't know how I was going to discuss it with you,' she continued, 'even attending the funeral didn't help.'

'How long will you be away?'

'I don't know. They want me to stay a few months – especially the winter months – and have also suggested some trips abroad. A kind of seeing the world before it's too late. I think it's a lovely idea and from there I just don't know. Anything could happen.'

He began to feel desperate.

'You can't do this. Not go and leave me – indefinitely.'

'Control yourself, Peter. You don't know what you're saying.'

'You just can't. You say it is better this way. Why? Are you afraid of people talking? I wanted to marry you.'

She paused in some obvious confusion.

He waited in a torment of suspense. Why do women love to make men suffer?

Eventually she said in a measured tone. 'Surely this is not the time, place or circumstance to discuss such matters.'

Castigating him in this manner she was actually blaming him for his generous offer. But she couldn't see it or not understand. Or was she merely taking a delight in torturing him?

'Did you really intend going away without letting me know?'

'If you hadn't come tonight I would have written to you.'

'Well, I'm here. You needn't write. You can tell me.' He was getting angry now. 'You can tell me what you really think. Penny, what are you afraid of?'

The calm way she'd approached the issue was enough to madden any descent man.

'I'm not afraid of anything. If you feel the same way when I return – if we both feel the same way – then I suppose...' She smiled grimly but with full control; yet she didn't hold the smile for long and she talked on to conceal its disappearance.

Yet he found it hard to listen. He had his own thoughts. Was she attempting to preserve her freedom and possibly do better by looking around in a different locality and especially in undertaking a lot of travel? What an excellent way of meeting people – especially people that mattered, the somebodies, the leisured classes? The world was open to big opportunities and much better than tired old Welwyn.

When he suggested she may not feel the same when and if she returned his words were bitter.

She shrugged. 'It would be foolish to decide things – now – tonight.'

'Why not?'

'Well after what you've been through.'

'You don't know the half of it, Penny.'

Her interest was rising. 'Don't I? What do you mean by that?'

'You didn't know that Karen knew about us.'

'No. I didn't. How awful. And how she must have suffered. I never dreamt she knew.'

'Now you can appreciate how I suffered to conceal things, the hypocrisy, the secret dealings, the planning...'

'And she knew and you say you suffered. My God she must have suffered,' she said with obvious derision. 'Both of us must have made her suffer. She just gave up. Peter I can't take much more of this – especially tonight.'

He stood up, placed the large envelope on the table and said to himself – perhaps she might forget. But he certainly would not. He realised his future would be afflicted with long arrears of remembrance. Just as Penny had described Karen's secret sufferings, along with bearing so bravely her protracted illness, so he would see her in years to come. He would have to pay the penalties attached to attempted oblivion and it would not help, perhaps only hinder, to have Penny pay with him. The burden would have to be his alone.

At last the inevitable words that were slowly coming to the surface came forth.

'I'd better be going now,' he said and left.

# 16

A strategy still had to be worked out. Besides his usual column in The Guardian Peter Smith would carry out an investigation into the benefits and businesses associated with cancer research. This was a tall order but since he'd been personally involved the stakes were high and, who knows, a major scoop for the paper could be achieved. He'd have to plan carefully and give little away, especially in the initial stages. He had been particularly pleasant to Dr. Nicholson before leaving Addenbrooke's.

His first port of call was the Q E 2 Hospital in Welwyn to interview Dr. Wilkinson. There was no difficulty in getting an appointment.

'I was sorry to hear about your dear wife,' Dr. Wilkinson said as he greeted him.

'Thank you, doctor. She's such a loss. The house is now an empty shell and as we have no children it is particularly hard.'

'I'm sure it does. A great tragedy. You must realise we could help her no longer here. She was rapidly getting worse.'

'Are you sure, doctor? Could it not have been one of the many relapses she had over the years?'

He shook his head emphatically. 'No, Mr. Smith. This was different. We were losing control. All the parameters were going off beam and we knew when we were defeated. You must accept one's professional judgement.'

'Oh, of course doctor, I do. There's no question of that.'

The next interview arranged by Peter Smith was in Landers' Pharmaceutical Company in Stevenage. Because he was special correspondent with the newspaper he was received profusely by their Public Relations Officer, Chris Neil. Peter Smith had already written to them and an appointment was arranged. He parked his car in the visitors' car-park alongside the impressive modern building. Then to Reception. Yes. He was expected and a languid debutante invited him into the visitors' lounge. This room was an eye opener and an elaborate front for the huge clamour of international affairs that went on behind the scenes; the effect was of a place where first things came first, second things never. All the outward show of efficiency was merely the padded glove over a mailed fist, the smile on the gladiator's face. The debutante remained in light conversation, offering coffee or anything else he desired.

Then a thin quiet man of about fifty appeared. He had a handsome, restless face and clear, fresh, trusting eyes – on a first impression at least.

'Please come this way, Mr. Smith,' he said holding out a welcoming hand. 'We received your letter and some of the details were a little disturbing. Let's make our way to my office where we won't be disturbed.'

They walked through corridors almost to the back of the building and for a moment Peter thought he heard barking although he could be mistaken.

'Please sit down, Mr. Smith,' said Chris Neil when they arrived. 'There are things we should discuss arising from your letter, which have caused disquiet in this organisation. The information you've obtained through the unfortunate death of your dear wife has been an illumination and shown how harmful this particular activity could be to the interests not only of this company but throughout the industry. This Dr. Nicholson could potentially undermine, or even destroy, the basis for one of the most lucrative generating businesses in the country -- after the oil industry.'

'I really don't understand how a simple drug could do this, Mr. Neil,' replied Peter Smith, eager to learn more.

John Smith's determination to find out more about the workings of drug companies intensified. It was a well known fact the industry was in the news on a regular basis in the sixties – and was mainly unfavourable.

Now years later he assumed that most of these practices had been cleared up. He was in for some surprises. With his incisive mind and devious techniques he set about finding out all he could about what went on behind the glossy surface. He attended a number of international conferences and got the feeling that meetings tended to start off with a certain bias and proceedings would move from preconceived ideas to preconceived conclusions – most times by subtle means which were hardly noticeable.

He kept a detailed account of agendas, procedures, voting tactics and the astute handling by a chairman of some potentially embarrassing questions. Regardless of this unfairness certain ugly truths became apparent. Amongst these he discovered an excessive degree of price-fixing between companies and collusive practices, illegally rigged quotes for government contracts, deliberately misleading information on drugs including misinformation about side effects, infiltration of the Health Department to sabotage contracts, bribery and corruption of certain high ranking people with 'honorariums'. And the list went on. His file became bigger and more alarming for a lay-man to read.

One official tried to reassure him that there was nothing amiss 'because in this business, most doctors know little about drugs, and if you treat them straight, they will trust you and you can get all the business you want – especially if you remember to treat them like Gods.'

# 17

Late one evening Dr. Nicholson was writing up charts in the nurses' station and he overheard a conversation between an old patient and his son. He recognised the voice of Graham Sinclair so he moved to the door, closed it to allow them privacy and went back to his desk.

Conversation continued inside the room as the son urged his father to recall a certain story.

'Sure, you'd not be interested in that kind of thing at all,' the father protested.

'I would. Of course I would,' the son said.

'Course you wouldn't. All you're interested in is medicine, books and music invented in hell to awaken the dead.'

'No dad. I'd really be interested. Go on and tell me what happened.'

The father gave a long sigh.

'Well. If you insist. But you'll have forgotten about it by this time tomorrow.'

'No I won't. I swear. Tell me what happened when you and Colm went off on Dutch leave and...'

The father lowered his voice.

'Will you be quiet about Dutch leave? Walls have ears, you know, and anyway this was years ago.'

'Go on dad. Tell me. My lips are sealed.'

'OK. As long as you promise to keep it to yourself.'

'I will. I promise.'

The father rested his head on the pillow.

'OK then.' He stared into the surrounding darkness. 'Colm and I decided to visit the local hostelry in Normandy – it was towards the end of the war. Nobody noticed as we slipped out through one of the side entrances of the camp.'

'And then?'

'Don't rush me lad. It's hard to remember details. It's so long ago.'

'Of course it is. But try your best.'

'Well we succeeded in giving the guard the slip. He was half-asleep anyway and off we went with not a care in the world. There was a lull in the distant gunfire as the night was very dark. But we had our lamps so we succeeded in staying on the road.'

'Lucky you. What happened then?'

'Well, we headed towards the village and both of us had a few bob, or was it dollars in our pockets? I can't remember. Anyway we made for the first pub. It was half-full with people –young and old who were mainly locals. We found an empty table and ordered a litre of their best *vin du pays* which arrived quickly once the dollars were seen and, you know, the thirst was awful. We polished it off in record time and ordered a second. Again it arrived in record time but we slowed down and had a good look around at the action. Most of the talk was in French – double Dutch to me and my pal – but in the next booth we could just about make out English in subdued tones. My friend, Colm, put his finger to his lips and looked at me crossly as I was happy to mumble on.'

'Could Colm hear what was going on?'

'We both could but it was difficult. Apparently, the three blokes next door were discussing final preparations for an attack on our barracks the following night.'

'My God, who were those yokels working for – the Germans or what?'

'I don't really know, except they had northern accents – you know Belfast or around it.'

'Yes. Easily recognisable. What were they saying?'

'They wanted to put a bomb near the arms dump in the camp, so there would be an almighty bang. We were shocked because we had our quarters near the dump and I didn't like to think of the consequences if they succeeded.'

'Go on. What happened next?'

'Colm and I continued in our finest French which was gibberish but I suppose it sounded genuine to our friends next door as they continued to talk.'

'What were they saying?'

'Don't hurry me son. It's a good while ago now and details are hazy.'

'OK dad. Take your time.'

'Well anyway it was about nine o'clock and we were just two local farmers chatting away in awful French with long pauses. It transpired that a group of them would gain access the way we got out – or were going to cut the wires – I can't remember. But they were determined to succeed in getting to the arms dump. They had maps and everything.'

'Who on earth are they Colm asked?'

'Can't you guess, you French git I said? They are a group of IRA wanting to inflict as much damage as possible on the enemy.'

'My God, what evil rotters and they might have ended up killing many and probably some of their own countrymen.'

'Sure. Sure. The length some people will go to. So details were finalised and they decided on a time for the event – midnight when all would be quiet and we would be asleep in our bunks near the dump.'

'What happened then?'

'We whispered to each other as one of them looked around the noisy room. And you know, son, I recognised him. He was one of our group – ugly bastard with pimply face. I won't forget that face for a long time.

'What then?'

'We played stum, and complained about the weather and the state of things.'

'So you were on to them and they didn't realise?'

'We sure were. We had to make it back to camp as quickly as possible.'

'To spill the beans?'

'Yes. But of course that would get us into trouble so we had to think of an excuse – or a course of action to carry out as Captain Rochford would put it.'

The father continued to explain that he and Colm left silently and returned to base unnoticed by the plotters. The intended action was immediately reported to the OC of the camp and an ambush was prepared. A reception party caught the group red handed and they were taken into custody.

At this point Dr. Nicholson turned the reading light on in the nurses' station as it was too dark to write properly. Immediately the son turned around, opened the door, saw Dr. Nicholson and realised he might have overheard some of the conversation. He looked at his father who had collapsed back on his pillow with eyes closed – hoping. He approached Dr. Nicholson and asked had he heard the conversation between himself and his father.

'No. Not really. I was too busy writing up notes,' Nicholson said.

'But you heard some of it?'

'Yes, but it doesn't concern me. Why should it? Your father strikes me as a very brave man. And he has plenty to be proud of.'

'Yes, he has.' The son hesitated. 'Ever since he has had a dreadful time – changing his identity, his address and his family have lived in dread -- for doing the right thing.'

Sinclair became aware again of Nicholson's slight Irish accent and politely asked him where he was from.

'Born and brought up in Mayo, near Westport and proud of it. Why do you ask?'

There was no reply. Sinclair was about to turn away when Nicholson said.

'Look, Graham. I'm a doctor. Nothing else. I've no interest in politics or revenge or whatever you might think.'

'You do realise Dr. Nicholson that my father has suffered for his life ever since, dreading a reprisal or some revenge and so far he feels safe.'

'I understand that completely. Incidentally, why is he in hospital if you don't mind my asking?'

'High blood pressure and his nerves...'

'Well this is the right place for him then.'

Sinclair looked troubled.

'Is it? Now that the greatest secret of his life is exposed to a stranger and a fellow Irishman at that I dread to think of the effect it might have on him.'

Nicholson looked at Sinclair intensely and said slowly.

'I will say nothing of what was said tonight. I give you my word and especially your dad. Nobody is going to come and get him – as you imply. I promise. I've always thought that bigotry develops as a rage in one who has no opinion.'

'I hope you're right. Just our luck to talk and you nearby. He'll be convinced they'll be after him now and especially with his borderline dementia as well as the blood pressure.'

'Again I promise I will say or do nothing and I always keep my word. Please don't worry.'

'It won't make any difference to my application to join your unit – will it?'

'Not in the slightest.'

With that the son returned to his father's bedside and reassured him that all was well.

His father was rambling, with eyes closed, saying. 'The mind is like a parachute – it only works when open, and anyone who has tried to survive poverty will know how expensive it is to be poor.'

He opened his eyes and looked sadly at his son.

'Oh Graham. They say you enter and exit this world alone, yet I felt more alone after your mother died than in the coming and now the going. I've tried but find it difficult to pray.'

Graham held his hand and rubbed it gently.

'Dad, words are not necessary. The mere wish is a prayer in itself. And your prayers are really valued by God especially at this critical time of your life. All will be well.'

But all was not well. Next day the father suffered a massive basilar stroke. The outlook was very poor. This was what he'd feared all his life. 'They' would eventually catch up on him and exact their punishment. It was unnecessary because Nature had done the job for them in a matter of hours.

He died three days later – for doing the right thing having suffered decades of hell waiting and waiting. Now the waiting had come to an end.

Was there any justice in the world asked John Nicholson of himself when he happened to be in the wrong place at the wrong time?

# 18

The diverse nature of cancer research soon became apparent to Peter Smith. There was a delicate balance between universities and the pharmaceutical companies; cooperation took place as long as the status quo continued with no interference from anyone to upset profit margins. The industry kept strictly within guidelines; each product tested on animal models and then sometimes on human volunteers after which it was released under special patent. The licence granted by government was the most valuable aspect of the process.

However, after taking legal advice from a friendly solicitor in the city and legal team on The Guardian he decided to test the water. He would make a formal complaint to the General Medical Council setting out the facts of his dear wife's death – as he saw it. He wrote telling them she had been ill for about seven years with a Non-Hodgkin's lymphoma and had been well managed by the oncology team in the Q E 2 Hospital in Welwyn.

As far as he knew everything was proceeding normally so he went on a week's conference in Bournemouth. It was during this time, he was told later, that his wife became seriously ill and for some reason was transferred to Addenbroooke's Hospital in Cambridge. He was not informed of this and no reason given.

When he returned from the conference word got through to him that his wife was in Cambridge where she was being treated with new experimental drugs that required the patient's written consent. He couldn't agree to this but apparently treatment went ahead without his or his wife's written consent. Some days later she died in ICU.

This was the evidence sent to the Council who gave it serious consideration. Dr. Nicholson was summoned before a committee to defend the case. Fortunately he had completed a complicated submission for the University Grants Committee that required a lot of scientific effort and a copy was made available to each member of the committee. The hearing took place with Dr. Nicholson present for a certain period and after searching and difficult questions the committee decided unanimously that there was no case to answer and the complaint was dismissed.

# 19

Peter Smith rushed into his office in The Guardian and sorted through a pile of correspondence. One letter caught his attention. It was from the General Medical Council. He opened it and rapidly read the contents.

After thanking him for his submission about the circumstances of his wife's death the ethics committee discussed the merits and any possible negligence. The discussion was detailed and exhaustive and as a result it was decided that Dr. Nicholson should be invited to a second meeting. He did attend and brought copies of a comprehensive report that was prepared for the University Grants Committee. There was a long and erudite explanation of all the experimental procedures carried out in the Oncology Department of the Hospital. A lot of the material was new and innovative. Members were impressed with his work and after searching questions, which he answered fully, it was concluded no evidence of negligence by Dr. Nicholson could be found and everything possible that could be done for Karen Smith had been done. On this basis, together with all the other findings, the committee came to the conclusion Dr. Nicholson had no case to answer and it was therefore dismissed. They trusted that Mr. Smith would accept their decision as a true and accurate account of the hearing into his complaint and expressed sincere regret over the sad loss of his wife. If he required further information the Council would be glad to offer any assistance.

Peter Smith threw the letter on the desk with contempt.

'What a load of rubbish,' he murmured, 'from a crowd of fossilised fools who don't live in the real world.'

Yes, they'd looked into the case with reversed binoculars and found Nicholson was carrying out research that perhaps they couldn't even understand but he must be a good chap because he'd written a blindingly complicated scientific document for the University Grants Committee – and, well, that was good enough for them. It jolly well had to be because there was no alternative source of evidence provided. This was not a court of law. Even though their avowed aim was to protect patients it was still a body of doctors sitting in judgement of other doctors. One could liken their deliberations to the police investigating the police for misconduct. That had to be the similarity. All professions worked against the laity.

Over the next few days he was so incensed he drew up a column for his paper on the unfairness of the General Medical Council being allowed to make judgements on misconduct of its own members – like a closed shop – with no public access to the findings and no accountability. The injustice of the situation should be made public.

It was a hard-hitting report, full of vindictive innuendo and initially the editor liked it. Subsequently he got cold feet. He discussed it with Peter Smith.

'You know, Smith, if you publish this you're going to shoot yourself in the foot for any future possible litigation.'

'I don't understand.'

'Well if you publish the findings of the GMC – no evidence – case dismissed. This will prejudice any future court proceedings and I get the feeling you are determined to take this to a higher court to get satisfaction and complete resolution. Just keep this article under wraps for the moment and talk to those who really have a vested interest in cancer research – the pharmaceutical companies.

They don't want to see their drugs being altered and tampered with so that their patents are no longer watertight. Maybe that's what Nicholson is at – to undermine the whole industry.'

'That could be a major scoop.'

'Yes, if it's true you could have a far bigger story that could have international implications.'

'You're right, sir,' Smith said triumphantly, 'changing patents slightly could relieve companies of huge profits and play havoc with the industry. Nicholson could be involved in this. The mind bogles at the thought. It could be horrendous and I never really considered that side. This could develop into a big news item for the paper and, of course a major protection for the industry. The GMC is just small fry in the overall picture.'

'OK. Smith. If you're serious in pursuing this legally the paper will back you but you should ask and get backing for legal costs from the relevant drug companies that feel they are going to lose out – Landers for example. Legal costs can be astronomical and this paper will not support you completely. Inscribed law is merely the web of spiders. It is sure to entrap the vulnerable and the poor, but would be shred in bits by the wealthy and powerful and also one guiding rule of lawyers is to create business for themselves. Get all the help you can – scientifically and legally – especially pro-animal research establishments like Landers and then get the best legal team together to defend the status quo. At first their attitude may be inertia. OK we're doing all right, leave us alone. We lead the world in chemotherapy with animal research. No need for extraneous experiments that could ruin a lot of what has been achieved. Then on further persuasion it could be argued that it might be a disaster for one of the great export earners in the country. A national disaster if Nicholson is allowed to continue.'

'My God, you've really unearthed a great news bombshell -- if it's true, sir.'

The editor raised a finger and looked over his glasses.

'A word of caution, Smith. Immense is the power of the press but not as immense as the power to suppress and also be careful of the wrath of a patient man.'

Smith gave a shrug of indifference.

Over the next few weeks intensive discussions took place between Peter Smith and Landers Pharmaceuticals. On several occasions Smith was conscious of animal houses that were highly protected on the premises but no comment was passed between the discussion parties. He was successful in getting agreement on certain strategies pertinent to the exact chemical nature of Landers' output of anticancer drugs and, if necessary, their own biochemists would be prepared to appear in court to support the principle of patency infringement that was sacrosanct to the company. Their existence depended on patents taken out on their products and other companies who were potentially at risk. This threat coming from Cambridge must be taken seriously. It was intolerable. Even government departments could be involved.

# 20

John Nicholson's in vitro work continued unabated in the Oncology Department at Addenbrooke's and almost every day a new and exciting development occurred – the result of long and extremely hard work.

One morning Gillian, the secretary, put through a call from Dr. James Wilkinson of the Q E 2 Hospital in Welwyn who was anxious to speak to him.

John grabbed the receiver.

'Hello, James. Nice to hear from you. Any news?'

'Yes, John. I felt I had to talk over a number of developments here. Several days ago I had a request from a Peter Smith, who is special correspondent with The Guardian. He asked if he could have an hour's interview at my convenience and I agreed. Not much use in saying no to these fellows – they have ways of talking to junior staff and things could get into a muddle.'

'What did he want?' John asked a little impatiently. 'I spoke to him after his wife died. Initially he was upset but calmed down rather quickly and became really charming. You know James, an obvious enemy can be an affliction, but a pretend friend is a disaster. I could guess where he was going.'

'With me he was full of questions and I got the feeling he was briefed by legal colleagues before our appointment. He didn't take many notes although the questions were written down. He'd one of those small tape recorders that look unobtrusive but are effective in picking up every nuance in conversation. He asked could he use it.'

'You agreed?'

'Yes. There was really no alternative.'

'And what did he want?' John repeated.

'He started by asking what protocols are normally adopted for new cancer patients admitted to The Q E 2. He wanted to concentrate on treatment regimens; what drugs were used, their names and the companies who manufactured them. As this was quite a big question I took a copy of The National Formulary and Mims, underlined items in red I thought relevant and he ended up with a lot of information on generic and trade names. I don't think I did anything wrong – this information is available to any investigative journalist if he really tries.'

'Yes, of course it is. But I take it he only concentrated on drugs given and nothing else.'

'Yes that's right and he did ask was this common practice in most hospitals throughout the UK and I had to say – yes of course it was. He also asked were all these drugs properly licensed and approved by the MRHA or whatever the regulatory body is? I said yes as far as I knew but he would have to ask other people. I couldn't be certain.'

'He seems to have done a lot of background work. I wonder what's on the agenda.'

'I wondered too. But probably more interesting to you John, he then started asking questions, in a very polite way, on my understanding about your use of anticancer drugs in patients. Are they the same drugs as every one else uses? Then why is your treatment regarded by some as a little, shall we say out of the ordinary, and indeed how can you claim better results than other centres doing the same work? I felt I was on a sticky wicket there so I said he should either speak directly to you or read the large number of publications that are coming out of your department.

They appear regularly in the British Medical Journal and The Lancet.'

'Strange you should say that James. I've had requests from a well known law firm in Lincoln's Inn Fields recently. I wonder is there a connection.'

'That's odd. Why has the legal profession suddenly begun to take an interest in the treatment of cancer? Most unlike them. That's usually left up to us medics to worry about. Unless there's an agenda. What do you think, John?'

'I don't know but I don't like what I hear. Did Smith ask any further questions?'

'Yes. There were two, although I don't see the relevance to what we are all trying to do like saving patients' lives. He asked was there any way that Dr. Nicholson could chemically alter a well-known brand name drug and change it slightly or moderate it so that it, the new drug, could fall out of patency protection and if so are these altered drugs still protected by licence from the MRHA or whatever the licensing body is? My reply was, I didn't know. He would have to find out for himself; although before he left I did ask him why he needed all this information. The only answer I got was "we'll see." And then he left with a smile.'

'OK, James I think you've told me everything. Thanks a lot,' John said with a frown. 'One thing to remember is damaged individuals can be dangerous. They realise they can survive.'

'Yes, of course. The whole interview upset me in different ways, John, and continues to upset me along with a number of so-called improvements that Mr. Butterworth insists on pushing through even though there is opposition. He's so arrogant – he's right and everyone else's wrong. He's like one of those deaf politicians who keep on answering questions no one has asked. I think the great fault of the system is that most bureaucrats seem to care more for routine than results. I find it terribly upsetting. And lastly John I feel guilty for getting you involved in this whole charade.'

# 21

In the next few days John Nicholson became aware of hurricane forces developing around him. One redeeming factor was the exhaustive report for the University Grants Committee that at least appeared to satisfy the General Medical Council.

He suspected that a battle was looming although things were vague. In spite of this he was convinced his work was reliable, dependable and safe as proved by trials with conventional agents and he'd defend his results anywhere.

It was at times like this that he longed for solace and comfort. There was none. Karita was returning to Stockholm with her material for the final stage of The Trilogy and they'd agreed the deadline would have to be reached before they got married. How sensible it all seemed at the time; now we'll do this, and then we'll do that – so neat and tidy. But what about human needs, yearnings and passions and having a shoulder to cry on?

The cottage was desolate that evening when he came home; every room was empty and, even though he knew it, he'd to walk through them all checking this and that.

He had his Mozart and Walkman so he could listen to them in the open countryside. That was it. The evening was not yet dead and curlews were calling and sun rays beckoning. The rain had stopped, night birds prepared for darkness and the earth felt as if it were quickening and waking in growth. He walked quickly along his favourite path and only wished he'd some companion -- even a faithful dog -- for company. Perhaps some time later in life. He was too busy now. Leaves were darkening behind the whitethorn trees, the blossom still wet with rain. Dark green leaves were luminous in the grey, hedges behind them overshadowed him like an old man, yet everything gave the impression of a new creation.

All this was a balm and together with Mozart's mathematical music he was more relaxed.

On return he phoned Karita in Stockholm. He said he needed to get away for a week or two -- the pressure was building up in Cambridge but did not mention any impending litigation. Karita, exhausted from her location trips, was reluctant to go on holidays again and would welcome a rest to finish her work. Tempers flared – who cared about whom? What's more important than John's career, Tom's career or their relationship? John slammed the phone down and regretted it immediately.

Next day a courier was sent from London to Addenbrooke's with a special envelope to be delivered personally to Dr. John Nicholson; in it was a writ for negligence leading to the unlawful killing of Mrs. Karen Smith and taken by her husband Mr. Peter Smith who was suing for damages. The case would be heard in two months time and further instructions would follow.

# 22

The late afternoon sun was loitering pleasantly in Welwyn Public Park as Dr. James Wilkinson sat on a wooden bench and gazed blankly at ducks swimming on the pond. He'd come here frequently to be alone and for peace and quiet – qualities that were in short supply at present.

After an hour of relative quiet a storm was brewing out west. Swirls of dust and leaves rose from footpaths and disturbed children at play. People around seemed to be a long way from him in time and space. He was occupied with thoughts. Others were doing their own thing; some sad, walking alone with eyes downcast and others sharing personal secrets and occasionally laughing, which seemed absurd – his mood was certainly not for laughing.

In spite of the sun calling in straggling beams it left behind amongst the trimmed bushes and trees a feeling of more at ease than almost anywhere else. He considered how things had become so difficult in his busy life. The span of his years reached back in all its ineptitude, made up of his petty strivings. It seemed such a short time ago when he started as a consultant oncologist. Everything ran smoothly, there was order in life and he'd complete control over the way clinics and patients were treated. His staff was happy, cooperative and work was a pleasure. If he required assistance from the hospital secretary or accountant it'd always be forthcoming with little or no questions.

Certainly no questionnaires to be filled out so that they could be considered by committees and the process would now take enormous lengths of time. He soon learned the huge administrative department under Mr. Butterworth was reluctant to give a direct 'no' but would usually ask for more information about whatever item he required. Answers would be supplied but for some reason they would not satisfy certain committees and eventually a 'no' would be the regretted response because the hospital had seriously gone over budget again and cutbacks would have to be made including ward closures, sacking of nurses especially those on agency work. Other ancillary staff would also be put on a part-time basis so the fundamental needs of managing such a big hospital complex could still be maintained to a satisfactory level.

When he asked about the needs of patients he was told they were always considered important -- but like everything else they were also subject to the same severe budget restraints the entire hospital staff had to put up with. The problems of management were far more complicated today than ever before and were becoming more difficult as time went by. Everyone would have to realise this, otherwise far greater sacrifices and restraints on hospital practice would have to be put in place. When he tried to find out where all the NHS money was going he found it almost impossible to get clear, concise answers. Most replies would entail a carefully written twenty or thirty page document full of invented words and jargon that he found difficult to understand – perhaps, he thought, that was the intention of the committee who put forward their 'progress reports', which were written not to inform the reader but safeguard the writer.

The most distressing aspect of the reorganisation was interference in clinical practice. Years ago he'd freedom to see patients without being questioned by unknown persons in human resources to please explain the necessity of this, that or the other. The limited time he was allowed to see patients was another new rule he objected to. This was a requirement for everyone without exception so that clinicians could reach targets that were arbitrarily set. Performance indicators were set up to measure and analyse everyone's work and were easily stored on computers for comparison studies. And computers can now tell us almost anything, he thought. They can answer everything. Yet they are answers we may not have asked, and maybe never wanted to anyway.

The extent of the reorganisation drove Dr. Wilkinson into fits of depression and he had major rows with Mr. Butterworth – some patients were extremely ill and required careful assessment that took time; others were quite well and did not require much time. No. Mr. Butterworth regarded all patients the same – no difference was allowed in the overall scheme of things. Otherwise the detailed planning arrangements with government departments whereby financial advisers could accurately allocate time schedules for all procedures and, as long as these were adhered to, then one could have a smooth running hospital with no problems whatsoever. A manager's dream.

But what about the sick patient who suddenly developed complications – no allowances for this as long as doctors did their jobs properly. It wasn't the manager's fault if a patient got an unforseen complication. That was a separate issue and would have to be discussed in detail in some other committee.

Poor Dr. Wilkinson – life was complicated by interference from people who appeared to make rules as they go along but genuinely didn't know what they were doing.

Now he was resting on a park bench with the air full of spring sunlight and with a future filled with transient beauty spilling over into eternity. Nothing is more precious than life, more sweat than time; nothing more comforting than the smell of cut grass, the soft voice of moving air and bird song.

His life seemed to be devoted to consoling others with the truth about themselves, obsessions that worried them, lusts that degraded them, foolish things that preyed on their minds. Now he was looking into his own soul and seeing reflections he did not want to see.

Suddenly he was petrified. His hands started to tremble and

a cold sweat broke out. A group of children became startled and moved away from the ashen faced man staring blindly across the quivering waters of the pond.

As a gentle breeze brushed across the lake, the colour of the water changed. From the west the sun tossed little golden discs on to the surface. If he looked carefully he could see them skip from one ripple to the next until they joined a band of shimmering light near the distant shore. The evening was preparing itself for a dazzling display.

He sat for a long time watching the light fade from the sky. He had a fear of sunsets now; they were so final and ultimate. The storm decided to travel elsewhere to annoy other people. Mist thickened and distant vistas took on a gentle blue haze. The first evening star appeared and gave him a sense of reality. At times like this Nature had a strange trick of listening and waiting and he felt invisible things were watching. A bent tree silhouetted against the sky looked like a crouching man and most things around him probably slept with one eye open. The breeze blew cold as misery and it was time to go home and to another day.

# 23

One afternoon Nicholson received a phone call from Dr. Wilkinson's wife, Eva. She told Gillian she'd like to talk to Dr. Nicholson as soon as possible because it concerned her husband – a good friend of Dr. Nicholson.

'Yes. Of course. I understand, Mrs. Wilkinson. I'll get him as soon as possible.'

'Thank you, Gillian. I do appreciate it.'

Gillian found Dr. Nicholson and he took the call.

'Hello, Eva,' he said cheerfully. 'Nice to hear from you. Hope all is well.'

'No John,' she replied flatly, 'all is not well. That's the reason I'm phoning. You and James have been friends for a long time and have a lot in common.' She found it difficult to continue. 'John, it's probably not right to unburden too many things over the phone – it would be wrong.'

'Yes,' he agreed softly. 'What have you in mind?'

'Well. If it's not too much trouble – and you can find the time – I'd like you to visit us in Welwyn. Sooner rather than later. Perhaps dinner, lunch or anything. I'm worried about James. He is going through a stressful time at work. He doesn't talk much about it but over the last eight to ten months he's become withdrawn and unhappy. There I've said more than enough already. Could you manage a visit, anytime you say and we'll fit it in? It would be so important to both of us.'

'Yes, Eva,' he said in a courteous voice. 'I would be delighted. I'm off next Saturday.'

'That'd be wonderful. James is also free on Saturday. Could I suggest lunch here in the house? You know it, of course, and would one o'clock be all right with you?'

'That'd be perfect. I'll look forward to it – and Eva, for what it's worth I think I've a rough idea what it's about.'

She gave a gentle laugh. 'Yes. I think you probably do. Incidentally, James does not know about this as yet – but I'm sure he'll be delighted to see you. He has great admiration for you and it's hard to get him to stop talking about you sometimes.'

'We mustn't let that get in the way. So let's leave things completely open.'

'John you're kind to do this. I know it will do James a world of good and relax awhile – a thing he's finding more difficult to do.'

John's car journey was pleasant and relaxed with Bach's music to accompany him. He thought, perhaps a little flippantly, why squander money on psychotherapy when one can listen to the B minor mass? He entered Welwyn and admired the geometry and uniformity of the place. James' and Eva's house was an exception to the overall plan as it stood back well away from the ash-shaded village street, in a semi-public way – neither with or against the neighbours – unique in its own way. Some might regard it as a demonstration of old-world standards of exclusiveness – but nothing extreme. You could take it or leave it. Elsewhere there was a candid exposure to the public where most of the residents enjoyed furtive glances through shuttered windows and laced curtains and where 'neighbourhood watch' was what it was intended to mean.

The Wilkinson's house faced the public with a difference, escaping the terrible bombings of the war and the demolition pencil of the town fathers when the new town was planned. Eva opened the door and gave him a kiss on the cheek and a hug. They entered the lounge where James sat in an easy chair at the fireplace. John had not seen him for several months and instantly noticed how changed he had become; the complexion was an unhealthy sallow and the eyes deep-set and sunken. John said to himself: Yes, here's a worried man – but just ignore it at the moment -- go along with the flow. He glanced at Eva and his silence conveyed his concern. Instantly she looked away. Again he gazed at James and his shock deepened. James' face had shrunken and he shuddered at the amount of weight he'd lost. The eyes were deep-set in his skull, the skin was dry and lips bloodless. It was an effort to stand and a stoop indicated he was under a heavy load. A visible tremor was in the outstretched hand.

In a thoughtless moment John let slip. 'James, you're ill.'

'Yes. I don't feel well,' he replied weakly. 'Would you walk with me in the garden?'

John preferred to sit beside the fire and talk but James insisted; there was urgency in his insistence. Maybe he wanted to be alone with John for a while. He took James' arm and slowly found himself falling into step, listening carefully.

John tried to be flippant as he inquired. 'How's the golf, James?'

'Oh it has had to take a back seat. There's never any time. So many dammed things get in the way.'

'Really. And you almost a pro. The last time I tried to play you plastered me.'

'Yes. We did have some good times. They were terrific and you were not too bad yourself, John.' A sparkle came into his eye. 'You know, John, I think advancing years with its so called serenity is only an excuse to explain our reduced capacity to react to joy and sorrow.'

John replied with a smile.

'It is preferable to be sixty years young than forty years old.'

When they returned Eva stood up and offered John a drink before lunch.

'No thanks, Eva. But go ahead yourselves.'

'How about you, James?' She asked.

'No thanks darling. If John's not having I'll skip it.'

'Fine,' she said. 'I'll pop into the kitchen and see how things are progressing so please excuse me. I'll leave you to catch up on the world outside – or whatever.'

'Thanks Eva.' John sat and looked closely at his friend. 'James, I've heard a rumour things aren't going too well at the Q E 2. Is that true?'

'It's a long story.' He was reluctant to speak mainly not to upset their guest with his personal problems.

'OK. It's long but surely there's a beginning,' John said eagerly, 'and we've all day if necessary. Please tell me what's bothering you. I want to help. You know, James, fear above all else can effectively rob us of our ability to reason and to act.'

James glanced around the room with unseeing eyes.

'The most appropriate words I can think of are interference and frustration. You work for the university John, and have an honorary consultant contract so you're protected to some extent from the inane stupidities of so called reforms. We NHS doctors get the full frontal attack, which is long term and getting worse. I'd suggest it's a matter of power and getting control of the system. When I started in the seventies procedures were straightforward, pleasant and gentlemanly. We doctors and nurses did medicine and surgery and were helped by a couple of administrators. Now it seems we've done a complete circle whereby a decreasing number of doctors and nurses are told what to do, or not do, when and what to prescribe or not prescribe by an army of government appointed bureaucrats who think only they can manage the entire NHS with little or no interference from the medical profession. As far as Butterworth is concerned I suppose it usually takes one mug to find a greater mug to praise him. If I'm lucky I get a definite "maybe." If he says he's making a "realistic decision" I know he's going to do something bad and tells others it's only his duty. As far as patients are concerned they're kept in the dark as much as possible because what they don't know they won't miss, unless, of course, some interfering young journalist acts as whistleblower.

Then the politicians and spin doctors are brought in to put out the flames and launch a pretty vehement blame-game that is usually aimed at doctors. Moral indignation is just dressed up jealousy.'

John nodded and continued to listen.

'The blame-game is going on all the time and the nearer the top the dirtier it gets.'

Raising a hand at this stage John said. 'I agree with what you say, James. I've seen it happen elsewhere and I don't like it. Remember our colleagues in general practice are also subjected to horrendous control problems, which differ in various places.'

'John, I'm afraid the medical profession is losing control of its own destiny and before long we'll be just automatons controlled by rule books and computers with dire punishments for anyone who does not fulfil targets and achieve performance indicators that measure everything we do. No wonder an increasing number of young doctors are emigrating or going private or leaving medicine altogether. The government has almost completely destroyed NHS dentistry; maybe they've the same agenda for medicine in the long term. John, answer me this – why must we be surrounded by people who appear to accept that sickness is self-inflicted or culpable, and that they have a preference for missiles than medicine? Also, I don't believe that suffering enhances character – happiness may occasionally – but suffering makes people bitter and spiteful.

John again raised a hand. 'Enough doom and gloom, James. We still have patients.'

'Oh, no we don't!' James said, his voice rising. 'That word patient is now politically incorrect. We all have clients. Yes, clients are the new people in beds and in waiting rooms and if the client is not satisfied he's encouraged to complain like in Tescos!'

'My God, James, what a scenario. It's like a great stranglehold on the health service that the public must be made aware of and if people realise how much wastage is going on with -- well -- your words " interference and frustration"-- we could actually go back to the good old days when patients were not clients and were properly treated with the best medicine and equipment available. That should be possible by redirecting the huge salaries of thousands of managers directly into the finest care of sick people – I mean patients.'

'Other countries don't seem to have the appalling problems we have,' James said with feeling. 'The system makes me sick. When an expert like you, John, works among us, you are sure to find that fools will form an alliance against you.'

John smiled and helped James to his feet.

After a pleasant lunch John glanced at his watch. Time to go and it was a long drive back to Cambridge.

'I must leave, but thanks for your hospitality. You were great Eva and I must take the two of you out again sometime.'

James hung his head and mumbled. 'Sorry to be such a poor host, John. There are few people around who can listen like you. I've got a weekend off to fill in forms about performance indicators and other questionnaires for all my staff going back some time. I've not been a good boy recently, so now I have to catch up. Think of me burning the midnight oil for the next few nights.'

John said nothing but raised his eyes to heaven.

Early next morning John Nicholson received an urgent phone call from Eva Wilkinson telling him, through floods of tears, that James had been found slumped across his desk surrounded by masses of paper. He must have died about 6.30 a.m. and the most likely cause was a massive heart attack.

*

After the funeral John returned to the house with Eva and a few friends. He was able to get some quiet moments with her in the kitchen, away from the buzz of conversation in the living room. He asked her to sit down and spoke in a caring voice.

'Leave Welwyn and its familiar surroundings for a time. Go somewhere different – not necessarily abroad – perhaps London or Edinburgh where there are lots of distractions. I can give you a note to friends and I assure you you'll be well looked after. I can also put you in touch with specialists who deal with problems you might have. Put yourself in their care and don't expect a sudden result.

This is a process of gradual adjustment to new circumstances. Go to theatres, galleries and anything that interests you.'

She looked sceptical for a while. 'Suppose I don't want to go. Suppose I just want to do nothing and let things take their course.'

John reflected on this, recalling episodes in his own life.

'There is something important I should tell you, Eva.' He was patient as he looked her straight in the eye. 'A man's dying is more the concern of the survivor than the affair of his own. And the saddest words of all are "it might have been" – like words at the graveside are those that have been left unsaid. Death is an integral and final part of our familiarity with love. It succeeds marriage as marriage follows friendship – or as winter follows autumn. To be lonely is no new thing. Sooner or later it is an unwelcome visitor. Families and friends die. That's inevitable. Husbands and lovers also die. We all suffer from Anno Domini. There is no escape. It's unavoidable and death is the greatest loneliness, which you've just faced. There's no medication for an instant cure. There is nothing that can be said to diminish the torment of what has happened. Unfortunately, grief is sometimes the penalty we must pay for love. It's something no one can escape.'

He stood up and gave a long sigh.

'If we attempt to do so by some means or other we recoil from it to a much darker and tortured place – ourselves. But if we face it end on and are aware there are others like us, if we reach out and try to console them and not limit it to ourselves, we'll probably find that overall we are better off. People can make a virtue out of forgetting – genuinely forgetting – and manage to regard memory as a monster. Sharing can be profoundly comforting. Tears can be shed because he is gone or a smile can appear because he has lived. Do you get the meaning of my rather clumsy way of giving advice, which mainly comes from personal experience?'

'Yes. You must be right.' She took out a handkerchief. 'I never guessed that grief could be so close to fear.'

John Nicholson had time to reflect on recent events as he drove back to Cambridge. During his time with Eva she'd shown him a note James tried to write before the grim reaper clutched him from their midst. It was unfinished but he was able to make out: I know I am very ill and may die before I am able to accomplish things that are dear to me. In spite of medical predictions I feel I am being hurried out of life and people expect too much of us. I am oppressed by the little time left and I wish to thank most sincerely... And there it ended in more ways than one.

News of James' death had come as a shock to many. It was of immense import and one of fate's irrevocable acts of vandalism. As a result all sorts of repercussions were to happen; yet there would be cover-ups and excuses offered by those whose only interest was getting a replacement; hopefully one who'd be more compliant to the smooth running of the Oncology Department, someone who'd know not to step out of line, do his job as laid down in the job-description booklet drawn up with great care by Mr. Butterworth and colleagues. In fact now was a good opportunity to insert a new list of rules and regulations for the next appointee.

As far as John was concerned James Wilkinson had filled his position in an incomparable manner, with great distinction and had a special place in the world. He was a medium for clear thinking, rigid conscience and a fine compassion for patients. He carried out his responsibilities to the best of his abilities and Eva was the flowering tree he'd planted in their midst – the source of his strength, which gave him rest and shade and the breath of dreams in its upper branches. Over the years his colleagues watched a man of his texture challenged by one stupid obstruction after another – ill health, misunderstanding and worst of all the new reforms in the NHS. He became submerged under the leaden embrace of impossible demands like a swimmer in a drowning grip. Then left alone he would show withered patches of regrowth.

Now he was gone.

# PART THREE

# 1

Press interest increases in the case of Smith v. Nicholson towards the end of the first week. Many witnesses are called for the plaintiff to give evidence and perhaps the most moving is that of Peter Smith. Called by the prosecuting council, Mr. Carton, he spends most of the day under close examination. The jury is exceptionally attentive. There is eloquence, poignancy and sadness in the things he says. Being a journalist he has a flair for words in convincing people of the rights of his case. He appeals to members to put themselves in his position -- with an ill wife whom he dearly loved being taken from their local hospital, unknown to him and without his permission and ending up in a frightening environment where her treatment is stopped – treatment she'd depended on for years. Then some untested, unlicensed medicine is given. She is used merely as a human guinea pig. To make matters worse the rules of the hospital are blatantly broken in that she does not give written consent for this treatment and, as we all know, she dies six weeks after being admitted to the Oncology Department run by Dr. Nicholson. Why did she die?

'I put it to you,' he says slowly with great feeling, 'she died because of negligence. She had done so well for seven years on accepted and licensed treatment – and then suddenly this. Her system couldn't take it and now she's gone.'

Mr. Carton raises a hand and says sympathetically.

'We understand what you've gone through Mr. Smith. No need to upset yourself any more.'

He faces the bench. 'No further questions, My Lord.'

The judge turns to the defence council, Mr. Beaufort, and asks if he wishes to cross-examine the witness. Mr. Beaufort hesitates then nods and stands up.

'Mr. Smith we have listened carefully to what you've had to say and it has become obvious to us that you had great affection, and even love , for your dear wife and, of course, we are all sorry for how things turned out.'

John Nicholson glances at him wondering what this is leading up to -- and a little puzzled.

Mr. Beaufort continues.

'Mr. Smith can you tell this court the nature of your profession please?'

'I'm an investigative journalist and have an appointment with The Guardian newspaper. I also free-lance for other publications.'

'I see. You must be a very busy man, Mr. Smith.'

'Yes. You could say that. I'm on the go most of the time.'

'And when your wife was moved from the Q E 2 Hospital in Welwyn to Addenbrooke's you were apparently on the go – as you put it.'

'Yes. I had to attend a week-long conference in Bournemouth. It's important for me to attend these meetings not only for professional reasons but also to be seen. I should also add that it is usually an enjoyable social occasion.'

'I can appreciate that and I'm sure you really did enjoy the meeting even though your wife was ill in the Q E 2 Hospital.'

'Oh, for goodness sake, I had looked after her for seven years.' Peter Smith explodes unexpectedly. 'I thought she would be safe for just one week while I attended a professional meeting that was very important for me to make new contacts and renew old friends.'

'Did you leave a forwarding telephone number?'

'No, but I told Karen where I was going. Was that not enough?'

'Well, Mr. Smith. I don't really know if it was enough. When an accuser points a finger at someone – in this case Dr. Nicholson – he must never forget that three are already pointing at him. And what I'm going to say next will have a strong bearing on the whole evidence presented to this court. You said you were at that week-long conference in Bournemouth, only to your sick wife. She was quite ill at the time and still you left her alone to battle with her illness. Then you apparently disappeared.'

'Disappeared! Sir, I don't understand what you are saying. I was at the journalists' meeting in Bournemouth'

'Mr. Smith, I would remind you that you are under oath to this court and I ask the question again were you at this journalists' conference?'

'Of course I was. I've already told you so. Are you deaf or something?'

Mr. Beaufort ignores the comment and continues.

'According to our evidence and on Dr. Nicholson's insistence an attempt was made to find you so that official hospital permission could be granted for the new treatment for your wife. Police inquiries revealed that you were not registered for the meeting and that you did not attend any of the daily symposia. If you didn't attend the meeting, Mr. Smith, where were you?'

Mr. Carton, for the plaintiff, jumps to his feet and says angrily.

'Objection, My Lord. This question is completely irrelevant. What does it matter if my client was or wasn't at the meeting or if he was in Timbuktu?'

The judge nods. 'Objection sustained. It is irrelevant where Mr. Smith happened to be.'

'But My Lord, with respect, it is relevant. Dr. Nicholson was obliged under hospital regulations to get written permission to proceed with his new protocol. The wife refused and the husband could not be found. I think we deserve at least an explanation from the husband.'

The judge hesitates a moment and then nods slowly.

'All right, Mr. Beaufort. I accept what you say. Objection overruled.'

'Now Mr. Smith you were being sought by the police and the address you left for contact was faulty. I think we should know where you were at this vital time.'

Peter Smith turns to the judge and says in a slow deliberate voice.

'With the greatest respect, My Lord, I think it is my business where I was at the time -- and nobody else.'

'Very well then,' the judge says coldly. 'We will have to ignore exactly where Mr. Smith was for the week of the conference and just assume he was somewhere else. As far as I can see it may or may not alter the nature of the charges against Dr. Nicholson. I think we should adjourn for today.'

# 2

The court reassembles precisely at ten o'clock next morning and Mr. Carton, council for the plaintiff, calls Mr. Chris Neil, public relations officer for Landers Pharmaceuticals in Stevenage to take the stand. He bows to the bench and then faces the witness with a pleasant smile.

'Now, Mr. Neil, please tell the court the exact nature of your work for Landers so that we can get a better understanding of the case so far.'

Mr. Neil is flattered and completely self-assured.

'My work involves the smooth running of the research programmes into finding the most suitable chemicals and drugs for many human illnesses, but we are particularly famous for developing a new batch of anticancer drugs that appear to be successful as measured in animal tests and then in human trials. We have published extensively on the subject.'

'Yes indeed your colleagues have and the resultant papers are well accepted worldwide.'

'I should add that for one new drug it takes two to four years testing on animal models at an average cost of £250,000 for each drug before it is ready for clinical trial on humans. This is just to give you an idea of the amount of preparation, time and expense involved in research and development in our company alone. And this is replicated in a lot of other pharmaceutical businesses throughout the country. It's an enormous industry costing huge sums of money to develop products and make sure that they are as safe and effective as we can possibly achieve.'

Mr. Carton turns to the judge and says. 'I've no further questions, My Lord.'

The judge faces the defence council, Mr Beaufort.

'Sir, would you like to cross-examine Mr. Neil?'

'Yes, I would like to ask a few short questions My Lord but I won't detain him long.'

He smiles at Mr. Neil.

'I would also like to thank you for coming here today to explain the work of Landers Pharmaceuticals and I'm sure Dr. Nicholson would also welcome your presence.'

Nicholson nods.

'Now I've been invited to ask you some questions on your research and development programmes at Landers. No doubt it is a wonderful company doing wonderful work and producing basic chemotherapeutic agents for cancer that have proved successful in a lot of cases.'

Chris Neil sits back and smiles benevolently.

'Thank you for your kind comments, Mr. Beaufort.'

'Yes, fine work indeed. Although there's something you said that worries me. It is this. For one product it takes two to four years to test it on animals and that the cost for that one product could be in the region of £250,000. I also understand that at any one time about 100,000 animals are housed in your laboratories. Is that true?'

Neil hesitates a little wondering where this is leading to.

'That would probably be correct.'

'And these are all healthy animals to begin with?'

'Of course. We don't experiment on sick animals.'

'That's logical, of course. You need fine healthy animals to experiment on, to inject poisons, bacteria and viruses into them. And lots of chemicals -- to make them sick.'

Mr. Carter immediately jumps to his feet.

'I object, My Lord. This is completely irrelevant to the present case and has no bearing on it whatsoever.'

'My Lord,' Mr. Beaufort says with obvious irritation. 'If I'm allowed to develop my point a little further I think you will see my comments are very relevant to what Dr. Nicholson is attempting to do.'

'Very good. Objection overruled. Carry on Mr. Beaufort.'

'Thank you, My Lord.' Again he turns to the witness. 'Well now, Mr. Neil, you have many animals that are deliberately made sick by various means – even giving them cancer – and then you try to cure them. But an awful lot are not cured, are they not? I have the statistics here and they are frightening. What do you do with the animals then?'

Mr. Neil remains silent.

Mr. Beaufort raises his voice.

'Do you kill them – all these sick animals, Mr. Neil?'

'I don't kill them – not personally.'

'All right then. They are killed and discarded. What an appalling amount of suffering and is it all necessary?'

Chris Neil gives a hopeless shrug.

'What else can we do?'

'According to Dr. Nicholson animals should be animals and left alone. When we decide to change their earliest development then we interfere with their cohesion and integrity and in so doing imply that Nature has no integrity. He says there is an alternative and a better way called in vitro testing of patients' malignant cells and their own normal cells using nanotechnology and with tissue engineering he is able to modify already existing drugs – fine tuning them if you like – so that they are more effective against malignant cells by a special target effect and less toxic on the patient's normal cells. He feels this is a more direct approach as opposed to your idea that mice are miniature men!'

Some members of the jury laugh aloud and the judge calls for order in court.

Mr. Neil nods and says he accepts what is just said but the

alternative method is a highly technical and skilled approach that can only be carried out in one or two pioneering centres like Cambridge. Very few other areas in the world have the know-how and protocols in operation.

Mr. Carton stands up again and loudly interrupts proceedings.

'I think, My Lord, we have heard enough nice talk on science and alternatives. I would like to remind the court that we have a serious allegation against Dr. Nicholson and that is of unlawful killing by negligence where an unlicensed drug was used. That, My Lord, is what we should be concentrating on with respect.'

'Yes you're right to bring us back to the fundamental issue, Mr. Carton. The court will adjourn and tomorrow we can hear the evidence of Dr. Nicholson.'

# 3

Next morning the court is full to capacity as Dr. Nicholson takes the witness stand. There is expectancy throughout the room as Mr. Carton, counsel for the prosecution, stands up after the witness is sworn in.

'Dr. Nicholson, for the sake of members of the jury I would like to remind them who you are. Please correct me if necessary. You are now Head of the Department of Experimental Pathology in the University of Cambridge and hold an additional position of Head of the Oncology Department at Addenbrooke's Hospital. Is that correct?'

'Yes. What you've said is correct.'

'Well then, my next question is relevant. Did you kill Karen Smith?'

A gasp goes through the packed court room. The judge bangs his gavel and shouts.

'Silence in court – please.'

John Nicholson stands his full height and says slowly and deliberately.

'No, Mr. Carton. I did not kill Mrs. Karen Smith.'

'But Dr. Nicholson, she is dead. You were in charge of her treatment. How did she die?'

'The immediate cause of death was a massive haemorrhage in the Circle of Willis in the anterior part of the brain – a well recognised weak area in female patients and rupture of vessels can happen after acute periods of stress. However, the underlying cause of death was a reticulum cell sarcoma of about seven years duration that appeared to be controlled with conventional chemotherapy until a few weeks before her death.'

'Yes, Dr. Nicholson, we know the sequence. She was transferred from the Q E 2 to your department, but her husband didn't know about it.'

'Correct. She was transferred to us in Addenbrooke's because Dr. Wilkinson believed we could handle her case better than the Q E 2 could. I agreed and went about setting up our usual protocol for new cancer patients with our tissue engineering techniques already explained to this court. I found the treatment Karen Smith was on did not deliver the maximum impact on target cells, that is the malignant cells, so I programmed an already known and accepted licensed drug in such a manner to target Karen's malignant cells in a more aggressive manner and cause less damage to normal cells.'

He pauses for emphasis, wipes his brow and continues. 'We did this successfully and Karen recovered significantly.'

'Ah, but I'm told you broke the rules of the hospital. For such a procedure to take place you must get written permission from the patient. Apparently, according to Mr. Peter Smith there is no evidence of written consent from his wife.'

'I had verbal consent from Karen Smith and this can be confirmed by Sister Mortimer in Addenbrooke's Hospital. If you want to call her as a witness she will confirm what I say.'

For a moment the judge sits in silence. He has a difficult case to resolve so he asks for a half-hour recession before he hears the final submissions for the plaintiff and the defence.

It is time for Mr. Carton to sum up for the prosecution.

'Ladies and gentlemen of the jury I do not intend to waste your time and attention on a lot of superfluous details – we've certainly had plenty of those over the last two weeks – although I did attempt on one occasion to keep the argument focused on the suit that the plaintiff has brought to this court. I'll remind you again. It claims that Dr. John Nicholson did deliberately and knowingly administer an unlicensed, unauthorised and illegal substance to a seriously ill patient in his own department. And also he failed to obtain the permission of the patient or of a close relative, that is her husband, before going ahead with the administration of this substance. Furthermore, he has admitted that this substance had not gone through any clinical trials and therefore the very nature of what he did was a foolhardy procedure. We know that soon after this substance was given to the patient she rapidly went downhill and died within a short time. The actual cause or causes of death are immediate brain haemorrhage – but I put it to you – she was stable and doing very well for seven years before going to Dr. Nicholson's department. On arrival her stabilising medication was stopped and experimental drugs were given to her against her will. She died soon after and I therefore ask you to bring in a verdict of negligence leading to unlawful killing of my poor client's wife who is now left alone in this world having suffered a tragic and perhaps avoidable loss of his dearest love.'

Mr. Carton pauses to allow time for the jury digest his comments.

'I, for one, would not like to be in the position of Karen Smith when the proposal to use a completely untested and unknown drug on me. I'm sure I would have refused to agree to its use – like she did – but circumstances militated against her and she lost her life. I therefore plead with you that the only logical, the only sensible verdict is negligence leading to unlawful killing. I rest my case.'

He reaches for a glass of water, bows to the bench and slowly sits down.

The judge nods his head and says.

'Thank you, Mr. Carton, for your elegant words. You certainly accomplished what you set out to do and that was to keep the facts simple and straight forward. This always pleases the court and makes life easy for all concerned.'

'Thank you, My Lord,' replies Mr. Carton as he closes his file and sits back in his chair with a smug smile. His assistant pats him on the shoulder and whispers some pleasantries.

The judge turns to face Mr. Beaufort.

'And now we come to the summing up for the defence. Would Mr. Beaufort like to come forward and address the jury?'

'Thank you, My Lord. I'm almost ready. I was just making a few scribbles on what Mr. Carton has said that I feel should be counterbalanced – but perhaps later.'

He turns to members of the jury and smiles.

'I also do not intend to blind you with details of science and technology. Firstly, in a court like this I think there is no need for it. We should deal in plain facts and weigh up the pros and cons of the arguments. And the main argument, as I see it, and as my learned colleague Mr. Carter bluntly asked – did Dr. Nicholson kill Karen Smith? My conclusion is emphatically no – he did not. Nor did he even attempt to do so. It never, ever entered his head. All he wanted to do was the opposite. He desperately wanted to save her life. It is well known that his ability and knowledge on human tissue engineering and the handling of nanoparticles is probably the best in the country. So you can see that simply means there is no one better to judge the wisdom of administering anticancer drugs in the right composition, in the right doses, at the right time – all worked out by enormous computer technology in his laboratories. We know from the late Dr. Wilkinson and colleagues in the Q E 2 Hospital that this expertise is not available at present anywhere else in the country so the most logical thing to do was to send her to this centre of excellence for what was the best treatment available anywhere.'

He shuffles through some papers on his desk as he continues.

'As regards unlicensed agents in the nineteen-eighties this is still an ill-defined and grey area and probably needs better regulation. I'm sure it will come in time. But for this particular case we are considering in this court – time was the essential ingredient that was missing for Karen Smith. She was rapidly getting worse and becoming confused. She could only give verbal consent and her husband for some unexplained reason could not be found to give consent. That life-giving or taking-away element – time – was missing. So Dr. Nicholson, with great courage and certainly not negligence, attempted, I would say heroically attempted, to save Karen Smith's life. He told me privately that if no action was taken at all she would almost certainly have died many days before she did. There was no doubt about that.'

Mr. Beaufort walks over to face the jury directly and opens his hands.

'So finally, ladies and gentlemen of the jury, you have before you a man who has worked tirelessly and unselfishly to produce wonder drugs against cancer by using new and revolutionary technology in human tissue engineering. He is also a man of great sympathy. The attempt he made to save the life of Karen Smith was a glorious act of compassion and humanity and most certainly, and in no way possible, was it an act of negligence. I therefore ask you to utterly reject the proposal wholeheartedly. There is absolutely no evidence for it.'

He turns and bows towards the bench.

'I rest my case.'

'Thank you, Mr. Beaufort,' the judge says. 'I'm glad you didn't go into great scientific detail, which is always a temptation in a case like this. I find you both have spoken well and that there is little need for me to give guidance one way or another. So I would ask members of the jury to discuss the case and come to a decision. And I will only accept a unanimous verdict. So please think carefully on what has been said. The court will now adjourn and await your deliberations.'

It takes two hours for the seven women and five men of the jury to come to their verdict. Word gets round the building that a decision is imminent and the court rapidly fills to hear the outcome. When everyone is assembled the judge bangs his gavel for silence and turns to the chief juror.

'Sir, have your members come to a decision in this case?'

'Yes, My Lord. We have.'

'And...'

'The unanimous verdict is not guilty as charged.'

The packed court room goes wild with excitement. Repeatedly the judge calls for calm and restraint and eventually peace and quiet reigns.

John and Karita hold hands on the defence side of the room hardly talking but handkerchiefs are used freely.

Finally the judge is able to speak when the furore settles down.

'Ladies and gentlemen of the jury I have to thank you for doing your job so expertly and precisely and also in such good time. In other cases we may have to wait for days. The rapidity of your decision – a mere two hours – and the unanimous verdict allows me to grants costs against the plaintiff and not allow an appeal to take place. I now have pleasure in dismissing the case against Dr. Nicholson. Case dismissed.'

# 4

The weeks following the trial are hectic for John Nicholson. Press, radio and television reporters hound him and members of his department. It's becoming too much. Fame and notoriety go hand in hand and interfere with his work.

Dusk descends as he reaches home. It's February so it settles at an early hour. He makes straight for the writing desk to confront a pile of papers that lie unchallenged. He sits deep in the chair away from the circle of light shed by the reading lamp and is surrounded by dimness permeating everywhere. Something is on his mind that prevents him from attacking the paper problems on the desk. He gazes into the increasing darkness.

Normally, at this hour, he would rest and rest completely. He had the knack of switching off from the problems of the day. Earlier the weight of the world would be on his shoulders and on reaching home he would cast this aside and assume the air of someone who'd spent the day in desultory pleasure and intended to end it in the enjoyment of domestic pursuits.

Not this evening. Karita, who stands just outside the circle of the lamp, is possessed by the gloom of the hearth from which an exploring flash occasionally lights up the expression on her face. She is worried.

John ignores her as his concentration is some distance away from the present. Eventually he speaks in a low soft voice but keeps his eyes fixed in the distance.

'Karita, the commotion of the media – it's just not real. It's driving me crazy and it's difficult to take anymore.'

'Of course, John. I understand entirely. You know I love you not just for your knowledge and ability but mainly for your heart that is so kind – so kind to me but also to hundreds of others and that's worth saving at all costs.'

He continues to look into the shadows for a while not wanting to disturb the ghosts that linger there. Then he explains his dilemma.

He has to get away. But where and with whom and for how long? He has a longing for Florence – hopefully with Karita. He asks her to come with him.

'Oh John, I'd love to go. It would be great but impossible just now. There's more work involved in finishing The Trilogy than I imagined and time is running out.'

'Then I'll go on my own. A week or ten days would be sufficient to get away from the maddening crowd. Would you mind Karita?'

'Not in the least. I can trust you. You're a loner at heart. You'll have freedom to do what you like. You must bring plenty of books -- but not textbooks. They're banned!'

'Of course. Thanks Karita. You're so understanding. I'll make the arrangements tomorrow. The sooner I get out of this bedlam the better.'

'Just leave a forwarding address and phone number. And don't worry about anything. Always remember true love is one soul living in two bodies.'

*

The early morning sun climbs the hill opposite John Nicholson's cottage. He slowly descends the steps from the front door. The remains of the night mist gradually dissolves as he crosses the road leading to the small lake on the opposite side. As he breasts the incline he's aware of the first waves of Wisteria over doorways and wooden fencing and the earliest signs of new foliage in sheltered corners of walls and buildings – he thinks what a beautiful way to start a day – one of those glorious English spring days that are found more frequently in memory than life.

The upper windows of the few houses nearby are tightly closed, for at this time the residents still sleep and he who's brave enough to venture forth at this ungodly hour wishes to enjoy what no wealth can buy later – the freedom of being alone.

Although John Nicholson's way of life, his elegant appearance and impeccable manners give the impression of someone of standing in the community, he's not unaware of the few pleasures of complete detachment from the outside world – and especially so after the recent legal battle that still haunts him.

On this occasion he finds a curious gratification in the thought of having the whole vista of landscape around the lake, the flowered border and not yet green grass moving into the water. He feels the morning is going to be brilliant when the fog lifts and every facet of the moving water intrigues him.

It has been an age since he'd time to stand and look without worrying what had to be done next. His flight from Gatwick is many hours ahead – so no hassle.

It is strange how he's growing to value these solitary occasions. In his earlier life he was dependent on the stimulation of over-zealous company and so-called feasts of reason and flow of soul. Now he watches out for and cherishes any chance to escape from the clambering crowd. Because of his ardent desire to have the world to himself he has cultivated the habit of early rising before he is caught up in the obtrusive social and hectic scientific world of research.

The flight from Gatwick to Florence is uneventful and John takes a yellow taxi to the Excelsior Hotel on the Piazza Santa Trinita near the Ponte Trinita over the Arno. He books for a night to give him a chance to look around. The hotel is ultra-modern, brash with severe lighting, which is unsettling. The bedroom is comfortable and he retires early.

He checks out next morning and immediately the magic of

Florence overcomes him – clattering sounds and bustle everywhere. And it's sunny again. He walks out of the hotel, crosses the road and leans over the wall gazing into the water flowing through the Arno Valley with its diaphanous light. The opaque green water is bordered by a pebbled shore with old flat houses in various colours on the opposite side. The skyline is irregular with pine and cypress against the morning light. To the right is the fragile Trinita Bridge and to the left the old Ponte Vecchio with its little shops. Beyond the bright sun shines on green pastures, trees and Tuscany.

He heads east, past the Piazzale d Uffizi and ends up in the Piazza Mentana where there's an interesting old house on the corner with red eaves and large green shutters. It is full of character. The name-plate on the wall reads 'Pension Mentana.'

Have you ever wondered at one of those old, very old, tall buildings in a narrow Italian street reaching for the sky competing with its neighbours with an old world elegance rarely found elsewhere? It has a long wooden shuttered front like a motionless mask, a mute face like a priest behind which are hidden the concealed facts of the confessional. Houses around them usually show off various actions and bustle shamelessly – almost boastfully – of life living close to the surface.

Isolated in its dark and mysterious frontage is the old Pension Mentana and is as impenetrable as the grave. Tall windows are like blind eyes, the door a closed mouth. Inside one could only guess; maybe it has a life of its own, sunshine, a fragrant myrtle and its own pulse of life through all the corridors, or a morbid silence where bats and other night- life have free access to their undisturbed homes and locked doors are rusty in their sockets from neglect and misuse and object with an eerie sound if opened.

John pushes the giant door, which resists at first, then yields to reveal a loggia with decaying frescoes and multiple terracotta vases hiding marble young maidens in various stages of undress. Sun saturates the small garden and quiet fountain, portico and grottoes. The air is lifeless and shaded archways afford a welcome escape from the glaring light.

He decides to find out what it's like inside and enters through more heavy wooden doors. He rings a bell at Reception and waits – and waits. The door is opened with difficulty by a shy peasant girl, who after a long scrutiny of him leaves him standing in the small antechamber. He hears her wooden clogs resound off the stone floor down a long corridor and after a delay beckons him to follow her. They pass through a more decorative room with lofty ceilings showing old fresco paintings. At the far end he is invited into a smaller room with the same stale colours about it but this time there are more signs of life. Faded tapestries hang from walls and it is almost impossible to see the illustrations on them.

'Have you any vacancies?' He asks pleasantly. 'For one person I mean.'

The bewildered girl does not reply but opens a door that leads into a heavily furnished lounge with a lot of frenetic grandeur about it. The girl disappears. After some time a tall elegant lady appears with dark brown Italian eyes.

'Buon giorno, signor.'

'Good morning madam,' John says slowly.

'Ah, English gentleman. I'm sorry to keep you waiting.'

'No Irish really. Irlandais.'

'Irish. That's even better. You are welcome. Please sit down.'

'Can I have a room?'

'Yes, of course. You may have a room.'

'What are your terms?'

'Forty pounds sterling per day. How long will you stay?'

'About ten days.'

'That includes full board – if you wish, breakfast, lunch and dinner and a large room, plenty of light and views of the river. Would you like to see it?'

'Yes. That would be nice.'

He's led up winding stairs to the top of the building. At last a bedroom with a large double bed, terracotta tiled floors over which sunlight streams and a lovely view of the river in both directions. He feels he could settle here. Its remoteness and old fashioned furniture are just what he needs.

The sky becomes dark and oppressive that afternoon, the wind turns cold and rain begins to fall steadily. He sits in the large bedroom and watches the green water flowing unaffected by the bustle of traffic on the opposite side. Afternoon tea is served about five o'clock in the lounge he had his first interview. Three women, possibly English, are drinking tea and eating biscuits. When he tastes the tea he declines to participate further and excuses himself. He makes an escape back to his far flung regions, lonely and abandoned, away and above everything. He welcomes the remoteness of the big old house. It is not dreary it is just plain indifferent – indifferent to cosiness, comfort and homeliness. There it stands – the showy overbright furniture ugly and foreign and trying to impress. For the present he likes the difference. Comfort is uppermost in his home in Cambridge with its large fireplace, low slung furniture and recliners designed for maximum comfort with thick rugs and heavy curtains adding to his kind of luxury. But here they tolerate the lack of fireplaces and heating. The gloomy coldness of the bedroom is a new experience. In a strange way he's glad to get away from warm extravagance. He likes the feel of cold in the air. He had too much of the cosy brightness of domesticity – he had to get away for peace and thinks he's found it.

A bell for dinner sounds well after eight o'clock – its official time. Down the winding steps and corridors he goes and finds the dining room where he is placed at a small table near the door. The women are already seated some distance away. Again they inspect him meticulously speaking in hushed voices. There is no need anyway as he cannot understand a word.

The only other men are the rather clumsy waiter, Pietro, and an aristocratic gentleman with an angry wife and child whose nanny is preoccupied with teaching the child good manners - to no avail. The meal is reasonably good and is served by the waiter and maidservant in a cheerful manner. There is a laissez-faire about the whole business – take it or leave it. Nothing really matters very much. Some incidents occur. The waiter lets a pile of forks fall noisily on the floor, the maidservant spills a carafe of water on an empty table and the child rejects the nanny and creates a minor uproar. At one time things seem to go out of control and he finds it amusing. He prefers this to the Excelsior where everything is super-efficient and smooth running whereas in the Pension Mentana nothing matters very much.

After dinner he returns to the sanctuary on the top floor where he feels he's in a fortified castle. Through the open windows the gentle sound of the Arno flows as it rustles over the stone covered shoreline. Traffic sounds deep and far away with sparkling lights on the opposite side. It is still raining outside so he retires to bed with its white gilded bedside tables and switches on a table lamp. He doesn't read for long as the weariness of sleepless nights creeps back on him. He surrenders to it and sleeps soundly.

# 5

John Nicholson wakes rapidly as February sunlight drifts across the terracotta floor and dazzles him as he turns on his pillow. A knock comes to the door.

'Who is it?' He croaks with the consciousness that creeps on the edge of sleep.

A female voice answers. 'Your breakfast, sir. You ordered.'

'Come in. It's not locked.'

The young girl with little English enters shyly and places the tray on the small table beside the window. She then throws open the shutters allowing in the glittering sun with all the crazy noises of the morning – the noise of a bustling Italian city, not a quiet cottage in Cambridge. The sound is so different -- like music to him, the complex orchestration to which his new life -- for a few days at least – will have to become accustomed to. It fills him with the joy of new discovery. Now his first days of no framework. He can do as he pleases and not have the hours hunting him like a pack of hounds.

The girl smiles and leaves the room. He sinks back on the pillow with a sigh of relief. Now for the joy of slowly getting out of bed, feeling the warm stone under his feet and going into the sparking sanctuary where his great porcelain bath offers a renovating temple of refreshment.

After a light breakfast he decides to get out and about as the rain has eased. Everything is wet; cars, carriages, wagons and people with black umbrellas move about their business. Then multiple ringing bells, especially the deep trembling of the Duomo, fill the air. All new as he walks past the rows of tall houses. He turns into the Piazzale d Uffizi and towards the end is confronted with the long slim Palazzo Vecchio going up to the sky. In the Piazza della Signoria he stands in awe. The stone paving in the empty square is shining wet and great buildings are gloomy and menacing. The front of the Palazzo, like a cliff face, soars up to the battlements and makes him dizzy. In front of this structure stands the great naked David, white and dripping wet and nearby more naked men in a huge fountain. It is the David statue standing in the position that Michelangelo chose is most impressive; he was the true master of Florence fully in keeping with the grim palace, its darkness in contrast to the whiteness of David. He stands back taking it all in; this grim square where so much happened in history. He is standing in one of the world's most significant centres – the Piazza della Signoria. He's a sense of having arrived, of reaching a centre point of highest human existence in spite of any worthless time barrier. Here is a scene like no other. Under an overcast sky, in this foreboding square are great naked men in the rain, supplemented by the fountain under the overseeing hawk – the Palazzo. It is a place that saw passion and fearlessness. Here men were at their zenith of achievement between the end of the old world and beginning of the new.

He feels a sense of revival, of fighting back. This is a city where men existed without apology and who had not to justify themselves. A rare finding nowadays. He's been told that to do the Uffizi Gallery justice it could take a day for each room. He's neither the time nor inclination – an afternoon is all he can spare so he studies the beautiful Botticelli's Birth of Venus, paintings of Raphael including a startling self-portrait, Carravagio's works and

Rubens' lovely painting of his wife who died one year after the portrait was finished.

Meanwhile back at the Pension Mentana the two ladies have finished lunch and decide to retire, for a while, to the roof garden as there's warmth in the air, which would be short lived. As they sit looking over the city one could see they are ladies of mature but well-cared for middle age. They lean on the veranda railings admiring the view. They can see long distances over roof tops. This is not their first visit to Florence; it is one of many yet these out-spread glories never fail to thrill.

The city seems trapped in a time-warp showing ancient buildings, the wonderful Duomo at its heart and the Arno winding gracefully through the centre and into distant Tuscany. What they observe now is no different from the Florence where great artists like Michelangelo, Botticelli and Leonardo produced their masterpieces, in which men saw and planned a new world and philosophers debated fundamental principles never before attempted, in which families of extreme wealth and extraordinary personalities fought for position and power. And men of vision saw and dreamt of things never thought of before.

In a previous life these ladies were theatre sisters in one of London's oldest hospital's, St. Bartholomew's, and had struck up a friendship ever since. Although both had married they'd lost their husbands not many months apart, which inevitably drew them closer. They liked to holiday in Florence especially at this time of year when the holiday season was almost asleep. They enjoyed the peace and quiet and each other's company. Almost always it would have to be the Pension Mentana and coincidentally each had the modest appendage of a salient daughter.

They discuss old times and how things have changed for young people. Daughters would certainly not be allowed out at night without chaperons and would be protected from goodness knows what – things too dreadful to speak of.

Christina Chadwick, the smaller of the two, says with a smile that brought colour to her pale face.

'What do our daughters think of us? Perhaps we've nothing better to do than knitting and reading all day.'

Anthea Hardcastle, the taller with a more pronounced colour, uses her dark eyebrows to great effect. She gives a careless laugh.

'Yes. That's the new order. That's what they think,' adding with a triumphant smile, 'the system has given us more freedom and time to ourselves. We can do what we like. That must be a welcome change.'

'Well, why can't we just stay here for a while?' Christina asks. She sits in a low comfortable chair in a sheltered corner and invites Anthea to join her.

'What a jolly good idea. This is probably the best view in the world. Why waste it?' She replies and sits with a sigh of contentment. The silence has a sort of diffused serenity enhanced by a perfect blue sky. Lunch is long finished and they have the entire veranda to themselves. 'It will always be my favourite sight. We were younger than our daughters when we first met here.'

'Of course,' Christina replies. 'We might have done worse. Who knows when they'll be back?'

Anthea is hard pressed for an answer.

'It could be a long wait, my dear. So let's get resigned to the new order of things. There's little we can do about it anyway.'

A pleasant hour or so is spent in silent thought on this elevated place over a gradually changing landscape.

John Nicholson has enough visual art for one afternoon and the evening is free. He sits in the gallery's coffee room for a well earned rest. A repetition of the previous evening's dinner in the Pension Mentana is not for him. Something a little more alive and exciting is needed. As he drinks his coffee he notices an old bearded gentleman that had probably seen better days looking at him – almost trying to catch his attention.

John raises a hand as a courteous gesture and as a result the man leans on his stick and ambles over to John's table.

'Mind if I join you for a while?' He asks in a gruff English upper class accent. 'I see you are alone.'

'No. I don't mind in the least,' John replies pulling a chair away from the table and ordering fresh coffee for his new friend.

The old man makes a contemptuous gesture and sits back with a sigh.

'A visitor here then?'

'Yes, but only for a few days. It's impossible to take everything in but what I've seen is glorious.'

'I see. I came from Oxford for two weeks and have stayed for ten years so far. The place is maddening. You think you know it – and next day you don't. Very frustrating indeed. I agree with you there's a work of art at every street corner. I used to pretend to be an artist but this place made that an impossible dream – impossible to imitate alone rather than originate. Perfection should not be used lightly but here – well I don't know...'

'You make me restless listening to you. I'd hoped to pack as much in four days and come away a happier man.'

The old gentleman raises his head and roars with laughter,

'You thought that? Of getting rid of your miseries in a short period. There are of course many ways of being miserable but I've learned there is only one way of being comfortable.'

John sits up and puts his elbows on the table.

'And what may I ask is that way?'

The other takes a long breath and looks into the distance.

'To stop running around everywhere looking for happiness. If you decide not to be happy there's no reason why you should have a fairly good time.'

'I'm sure that's good advice,' John agrees.

'In addition, young man, you'll also learn that the only reward of virtue is virtue itself, the only way to have a friend is to be a friend.'

John smiles and replies.

'You're quite a philosopher. Aren't you?'

'Used to teach it at Oxford but found it was a lost cause. Nobody listened. They knew best. So I took early retirement and here I am lecturing you.'

'I'm really enjoying our conversation. It's a relief not to be constantly struggling to make oneself understood if I wanted to buy some chocolate, for example.'

'Ah, chocolate and ice cream. Lovely.' The old man leans forward and lowers his voice. 'I've also learned that a crucial part of happiness is to be without some of the things you dearly want – sounds strange but it's true.'

'No. It makes good sense. Satisfaction brings boredom and there's nothing left.'

They mumble on about world affairs and then the old man looks at his watch and suddenly stands up.

'Goodness. I'd best be off. There's a lecture on Giotto at five o'clock and I promised myself I wouldn't miss it. I don't know enough about him.'

'Off you go then. It was great to talk to you.'

With that, the old gentleman limps away full of enthusiasm towards a lecture hall.

John walks along the Lungano beside the river – no restaurants there. Approaching him on their evening stroll are a pleasant looking Italian couple taking the air. He's nothing to lose so he stops and asks politely.

'Excuse me please. I'm a stranger in Florence. First visit in fact. I wonder is there a nice restaurant around here that you could recommend. They all look the same to me.'

The gentleman puts his hand forward. 'You're English. Of course, we'd like to help.'

John remains silent on the English jibe.

'We also are visitors to Florence. We are on our way back to Palermo in Sicily and decided to stay here tonight. My wife spent a week on a health farm near Lake Como up north. You know it?'

'Yes. I've heard of it. But never been there.'

'What a pity. Really beautiful area. Full of health spas and places to get rid of stress.'

John is surprised at their command of English. Immediately the gentleman hands him his card which reads: Professor Mario

Amaldi, Department of Animal Anatomy, University of Palermo, Sicily. It is in Italian on the other side.

John produces his card. 'Glad to meet you, professor. This is quiet a coincidence. Two academics.'

'Yes. A coincidence indeed. Please Professor Nicholson we would be honoured if you joined us for dinner in a little restaurant up in the hills behind Fiesole – a lovely location and they are good friends of ours. Would you do the honour of joining us for a simple Italian evening high in the mountains?'

'Well. I don't know. I thought I'd get something simple around here.'

'My friend Professor John. If you come with us you will have a night to remember even though the restaurant is a few kilometres away. I'll drive you there and back and we'll have a great time. Please say yes. Isn't that right Gina?'

His lovely wife smiles appealingly. There seems no alternative but say yes. So he gets into the back of the BMW 700 series that speeds off in the direction of the Fiesole Hills late that Friday evening in February.

# 6

The journey takes longer than expected; at one stage John becomes concerned but soon passes as Professor Amaldi says.

'Just five minutes more. We passed the sign back there.'

'Good,' John replies relieved.

On arrival there's a great welcome for the Amaldis – old friends meeting old friends – and now a famous new one from Cambridge in England. How delightful! They are ushered into a semi-private room of the restaurant, which is full of antiques, statuary and paintings and the whole layout impressive.

Gina excuses herself to tidy up and change. John and Giuseppe remain in the lounge.

'So it's your first visit to Florence, John?'

'Yes, and what an experience,' he replies.

'We always stop over on our trips to and from the north. There's no better place and Gina loves it. Ah here she comes refreshed and dressed for the occasion.'

John is startled when Gina returns. She's like someone else. Elegant she is in a revealing gown of thin red velvet with gold attachments down the sides. It outlines her shoulders, breasts, arms and legs – really modern. Round her neck sits a choker of pearls. Her dark hair is split in the middle, short and straight, pulled down over the brow in a fringe that highlights the large vivacious eyes. The make-up has a touch of exaggeration – intentional of course. He thinks her wonderful and a little overwhelming. A lady not to be taken lightly.

She looks around and says with a smile.

'Giuseppe is coming with the cocktails. I hope you'll like them.'

'I'm sure I will.'

'What did you do today, Professor John?' She says sitting down opposite him. She speaks in a strange husky voice. He is acutely aware of her firm legs thrust forward covered in red velvet, with its glistening effect and feels uncertain of himself. Never knew a woman could exercise such power over him. It's a bare naked force, although hidden, and just for a moment he cannot cope with it.

Giuseppe comes to the rescue handing him a glass. He takes it silently, so Giuseppe continues.

'Yes, John, what did you do today?'

'I went to the Piazza della Signoria and was amazed at the masculinity of the place and, forgive me, also the beauty of it.'

'No need for forgiveness, Dr. John.' It's Gina talking. 'I'm sure it's no harm for a man to admire men as long as it is kept in its proper place.' She smiles enigmatically. 'You do know what I mean, don't you?'

'Yes. I think I do.'

'Well. I mean as long as you leave most of your admiration for us women – then things should run smoothly as nature intended. Isn't that right, Giuseppe?'

Giuseppe is taken off guard and he splutters into his drink.

'Yes. If you say so dear. You're always right.'

'Thank you, Giuseppe. Now John you were saying you saw men in the Piazza. Men you admired. Did you see anything else?'

'Well of course. I went to the Uffizi Gallery.'

'Ah. That's much better.' She approved with a touch on his hand.

Giuseppe asked. 'And what did you think of it?'

'Beautiful building. But too much to see.' John said with a helpless gesture.

'What pictures did you particularly like?' It was Gina again.

'I liked the Botticelli's Venus in the Shell.'

'Was she pretty to you?' She asks rather pointedly.

'No. Not beautiful. But I liked her body. Better still I loved the atmosphere of the picture, the seascape effect and the fresh air.'

'What about her face? Did you find it attractive?' Gina persists with an exquisite smile.

'Difficult to say. A bit too innocent looking,' John said flatly.

'Maybe she was putting on the innocent bit and really she was a woman of the world,'

Gina suggested.

'Oh, no. I don't agree,' Giuseppe objected. 'She's the real Venus, wistful and innocent, which makes her a true modern Venus. That's why she appeals to so many people. A truly universal appeal.'

'Do you find any charm in Botticelli's Venus John?' Gina is now pressing the point home. Maybe she has her own reasons.

'I do not feel any charm from her. She has no appeal to me. It's only the whole picture – a wonderful work.'

'Would you agree she exhibits sham innocence and there's a lot going on behind the scenes?'

'I don't have much belief in innocence anymore.' John says and regrets it immediately.

Without realising it he's staring at Gina and she is looking away, yet knowing he's watching. Then she looks at him, with a slow dark smile full of understanding. He is still confused; maybe it is the country air, the wine, the strangeness of things. With the hypnotic look she gives he feels certain bonds of restraint melting away. His eyes remain fixed but he tries to smile back at her. Yet he is terrified.

# 7

They go into dinner. The room is small and intimate and lit by candlelight. Gentle Italian music flows with the sound of water issuing from somewhere – difficult to know the source. The table is round and Gina sits on John's right, Giuseppe on the left and flowers rest in the middle.

Immediately the owner produces three enormous menus. The Amalfis hardly look at them. They discuss a multitude of items in Italian – way over John's head. He merely sits back and smiles, hoping all necessary decisions would be made.

A starter is ordered – exquisite and expensive – with a light Moselle. All the while John is aware of Gina's graceful bare arms, rounded bosom, the pearl choker and the slightly obvious make-up. Something deep moves in him and he feels uncomfortable. He knows nothing about her – hardly speaks Italian – but body language says volumes. There is some loss of inner control that would normally hold him back. He sits there looking at her – she talks to him usually with eyes averted.

But really she says little. Giuseppe is the one who speaks straight from the hip. To John his manners are superb and he likes him in every way.

'So John, you are into cancer research in a special way. From what you say this could be very important. I'm a humble vet working in a southern university where we love animals. It's my whole life. Is there any way your work could help treat and may cure sick animals? We see many poor creatures suffering illnesses, but they die, yet humans are cured. It makes me sad to think they are our modern day slaves.'

'Well said, Giuseppe. I've not heard it said so eloquently and coming from an academic veterinary professor your remarks carry weight.'

Giuseppe looks surprised.

'Mamma mia. Not where I live. We are a poor community and anything I say can be ridiculed by the bosses.' He touches his nose and adds. 'You know what I mean?'

'Yes. I think I do.'

'In that case my voice will carry no further than the wall in front of me. But talking to you tonight it's almost like an act of providence.' He turns to his wife. 'Isn't that right, Gina?'

She nods vigorously and smiles at John.

'I hope I got a message to you about the terrible suffering of animals here in the Italian scene, but of course it is pan-European and probably getting worse.'

John looks at him thoughtfully. 'It's definitely getting worse. You must remember I'm not involved in the global issue of animal abuse in abattoirs and transport. I'm concentrating on replacing animals in laboratories by cell lines of patient samples.'

'Of course. Of course. I completely understand. And this is probably where the whole revulsion of animal abuse can start. You really are in a privileged position.'

Gina says little during the discussion. The husband talks most and John welcomes it. They both have a mutual interest. Her silence is almost remote but in spite of this John feels an attraction towards her. It is wrong. He knows it. But it is wrong.

The second wine is a rich light Bordeaux, a classic, soft and beautiful. She drinks it with great pleasure as if she understands its effects on others – the feeling is certainly infectious.

The dessert is tiramisu. John is shattered at the end of the meal and feels he's enjoyed himself so much he's no longer fully in control. Although his outward appearance is of a perfect gentleman, he's perhaps a little slow in putting on his coat and gloves.

They leave the restaurant to a roar of goodbyes and other appropriate sayings and he is driven back to Florence.

# 8

Giuseppe and Gina suggest leaving John directly home to his lodgings but he prefers to be left at the Piazza della Signoria – his favourite place – and he'd walk the rest of the way to the Pension Mentana. They reluctantly agree but ask him to be careful.

It is about ten o'clock and he is struck by the deserted feeling of the city. Before passing down the Piazzala d Uffizi he notices some men loitering around the David statue examining something in the dark. They form a weird looking group and it's obvious they don't want to be seen.

Instinct makes him move quickly, out onto the lungarno and then up one of the side streets, the Vin de Benci and left into the Via de Vagellai. As he moves along one of these dark streets there's a rushing behind him; three men are running after him. He hides in one of the arched doorways. They find and attack, knocking him from side to side and giving orders in hushed Italian. He struggles for a time but is held rigidly by one man while the others search his pockets. They take his wallet, watch and other small valuables. They then push him aside and rapidly disappear.

He'd been robbed – luckily no real violence was used and his passport and air tickets were safely locked away in the hotel. At least he is thankful for that. He struggles back home.

Dinner has been served a little later than eight o'clock in the Pension Montana. Both mature ladies appear at their usual table with easy conversation tempered with some anxiety about the return of their respective daughters. When dinner is finished there are no developments on this front so they agree to sit in the over furnished lounge to while away the waiting hours.

Nine o'clock comes and goes. Ten o'clock comes and goes.

They continue to wait for what seems like a long time – in silence. Christina Chadwick sits quite still, her eyes fixed on the huge fireplace with the marble carvings on each side. Andrea Hardcastle has stopped fidgeting in her voluminous handbag and also sinks into meditation. They have not had the occasion to be silent for so long before and Christina seems embarrassed by the silence. She does not know what to do.

Suddenly the front door opens and John Nicholson staggers across the hallway making for the stairs. His appearance is dishevelled and he seems uncertain of his movements.

The two ladies look at each other with knowing glances, which change to anxiety as Nicholson takes out a handkerchief and wipes his forehead. There is blood on it. And he wipes it again.

'Has he been in an accident?' They whisper and are tempted to approach him but he disappears rapidly upstairs with obvious difficulty. Again they look at each other – wondering. Should they leave things and mind their own business or investigate whether this poor gentleman needs assistance? After all both are trained nurses and well versed in handling patients who have experienced difficulties of one kind or another. In a state of agitation and indecision they sit in the lounge for twenty minutes or so. What should they do? What will they do?

'It's my own fault,' John Nicholson says to himself when he finally reaches his room. 'I should have been more careful and tried avoid these things. The Amalfis were right when they offered to drive me home at this time of night. I should have been on my guard. I jolly well deserve what I got.'

He's ashamed at being so foolish relying on the benevolence of others. He had it coming. How could he have been so thoughtless? He'd paid for his carelessness. Fortunately his travellers' cheques can be cancelled and reimbursed.

Never again will he be foolish and trusting. He is a fool, an utter fool. He's learned a lesson not to repeat such stupidity. To be alone is the measure of a man, he thought, and in so doing he must try to forget. What horrible things our memories like to harbour – the bad and upsetting – the lovely things are usually stored in diaries if we are lucky. Memory can be like a monster. I forget but it never does. It stores things away, hiding them and recalls them for no reason. Then I think I have memory -- no I don't. It has me!

He'll never again expose himself to any kind of mugging – not just physical mugging in a back street. There are far more subtle types of muggings in his professional life; he'd already experienced some of them. This incident highlights the pitfalls he must face on his return to Cambridge. Up to now he's been an innocent bystander until the recent court case. The real lesson of human malevolence had come home tonight – open straightforward robbery. Everything valuable – taken, gone. Just left to cope on one's own. Who cares? No one. To hell with you. No one is robbed unless he incites a robber – may not be a truism but lurking behind is the truth. In the same way no man is murdered unless he attracts a murderer.

He steels himself for future attitudes and resolutions. Now things must change. After this episode he'd surround himself with a mental bodyguard to protect his vulnerability. No matter what circumstance, whether it is passion or love, excitement or confusion, somewhere in the corner of his mind, asleep or awake, will be this sentinel to protect him. No matter what. Never to abandon him for an instant.

There is a gentle knock on the door. Who could it be at this hour? He ignores it for a minute or so. Again a more persistent knock along with an inaudible lady's voice.

'Please go away,' he says loudly. 'I don't need anything.'

The voice is louder and more persistent. 'We want to make sure you are all right. Please open the door. Please do.'

He makes a guess. It's likely to be the two ladies staying at the Pension. They are probably harmless, so he relents and opens the door.

'Oh good, Mr. Nicholson.' The taller lady says as she enters with her companion. 'We were sitting in the lounge waiting for our daughters to return and we noticed you entering the hotel. You looked dreadful as you disappeared up to your room.'

He looked puzzled. 'How did you know my name?'

'We inquired at Reception and they kindly gave us the room number.'

'I see.'

'We thought you might have had an accident or something similar.'

He closes the door and the ladies sit down.

'We are both trained nurses, in fact theatre sisters in Bart's in London, and when we saw you wipe blood from your forehead we were worried about your welfare.'

'How kind. I appreciate your concern.'

Then it all comes out – well most of it. The evening out, the walk from the Piazza della Signoria, the attack by three thugs and the robbery. How he managed to struggle back to the Pension and wished to retire immediately.

'How awful, Mr. Nicholson.' It's still the taller lady speaking. 'There seems to be a cut or wound on your forehead and your clothes are torn. Would you allow us to examine the damage? Remember we're both qualified nurses from London.'

'Of course. I'd welcome that. Please go ahead,' he agrees feeling relieved.

'We'll start by removing your jacket and shirt and ask you to lie back on the couch with your head on some pillows. Christina please help him while I wash my hands and get some towels from the bathroom.'

John does as he is told and soon discovers he's having a full physical examination with both ladies mumbling things to each other. Eventually, they ask him to sit up and look him straight in the eye.

'Well, Mr. Nicholson. As far as we can see there are no bones broken, no major injuries or abdominal wounds. There's just a small cut on your forehead and a dressing will soon fix it.'

'That's a relief. I was lucky there was little damage except the loss of money and valuables.'

Anthea Hardcastle nods and smiles broadly.

'Yes. They're replaceable and not that important if the money was only in cheques.'

'That's right.'

'Do you want to report it to the police?' She asks. 'We can help you with procedures if you wish.'

'I don't think that will be necessary tonight. It can wait until tomorrow. I'm due to fly home to Cambridge in a few days – in fact sooner if I can get a flight.'

'Cambridge. How lovely. We've visited there several times and it's a glorious place.'

Next he told them about himself, his work and other details.

'Dr. Nicholson, we never realised we'd such a distinguished visitor in our midst.' Anthea Hardcastle beams. 'It's a real pleasure to meet you, but of course, not in this way.'

Christina Chadwick has to have her say and selects her words carefully.

'Yes. You're right Anthea. Anything more serious and we would have taken you to the hospital straight away.'

'No. That won't be necessary. Thank you,' he says gratefully.

Anthea Hardcastle takes control again. 'Could I suggest we dress the wound now, get you into bed and take it in turns to check you through the night – just in case? No point in giving sedation with a head wound even though it looks superficial and harmless. Don't you agree?'

'Yes.'

'A mild pain killer and warm bed with professional observation is probably all you need.'

John feels embarrassed but agrees to cooperate. After all he has little choice. The wound is dressed expertly and he's prepared for bed with intermittent observation arranged by his two neighbours. He has a peaceful night.

Next morning things are back to normal. Police are informed with as many details as he can remember and the traveller's cheques are replaced. He takes the rest of the day off and stays in bed. All arrangements are made through the good services of Mrs. Hardcastle and Mrs. Chadwick whose daughters arrived home safely at 3a.m. after a glorious night out in a super nightclub.

It only remains for John Nicholson to collect himself together and arrange an earlier flight home. He's seen enough of Florence to last a long time.

Before he leaves the Pension Mentana he invites his two administering angels to dinner in a top class restaurant along with presenting them with something precious to remember their good Samaritan deed. Finally, if ever again in Cambridge they must look him up and they would be treated like royalty. The evening ended on a happy note.

Three days later John Nicholson returns to Cambridge a changed man.

# 9

On his return from Florence John Nicholson discovers that publicity of the court case is still raging in the media over animal versus in vitro testing. A hornet's nest has opened and polarization into opposite camps occurs – the pro and anti-animal groups; the former is by far the best organised with large funds behind it from the pharmaceutical industry -- almost a David and Goliath situation. Someone wrote Dr. Nicholson had won a case, a significant one at that, but what about future cases that could be launched against him?

The battle lines are drawn.

One morning Graham Sinclair phones John Nicholson requesting an appointment – he would drive from London to Cambridge. John agrees and a date is decided. Lunch is booked and they could probably retire to the office afterwards.

Graham has changed a lot. He is now a sturdy straw-haired man of thirty-four with a firm mouth and determined manner. Gone are the nervousness and uncertainty of his last days in the Q E 2. He briefly fills in his life so far as a mature student in London University. The course and content were superb and he'd learned a lot. The final examinations are coming up soon and a lot of preparation is required – and, of course, Helen his new wife is an enormous help.

'Glad to hear these things, Graham,' John says as they walk by the river after lunch. 'Let's sit for a while. It's so peaceful and away from the hustle and bustle of office life.'

'Good idea and the weather's great,' Graham replies. 'I suppose you know why I asked to see you, Dr. Nicholson.'

'I can guess.'

'Well, with luck and hard work I'll have my Master's Degree and I've already passed the FRCS, so obviously I'll be looking for a job that requires these qualities.'

'I understand. And I've always kept you in mind if anything turns up. Unfortunately there is nothing suitable at present but who knows? Things can change.'

'I'm aware of that and am prepared to wait. It has been a long two years. What's another six months or so? No, what I really wanted to ask – are there anything special, or any suitable quality candidates should have if a position arises? In other words how does one prepare oneself besides a piece of paper with a few signatures on it?'

John laughs and looks at a bunch of students racing their punts.

'You're absolutely right to be thinking around the point of application. Lots of dos and don'ts.'

'Yes, that's the kind of thing.'

'If and when an interview comes up I always like to hear the word enthusiasm. So when asked what you can offer, perhaps start with that. You see the panel will already know all about your qualifications and experience, publications and references from your CV. So they have nothing left but find out about you, the person, your views, your opinions and hopes for the future amongst other things.'

John throws a stone into the water and watches the ripples pan out as he continues.

'Some panel members try to catch you out with difficult questions to which there is no real answer. I should know Graham!

So you always must have an opinion, a mind of your own, because detachment, in other words "I don't know," energises the aggressor and not the victim. Remember silence is the greatest manifestation of contempt and the most spiteful lies can be told in silence. Always try to say something and make it simple. That is an acceptable technique. So have opinions on lots of relevant things and express them clearly. Do not worry about your opinions being regarded as a bit eccentric. If you believe in them that's all that counts because many old opinions are now accepted as normal.'

'Yes, before I left the Q E 2 I was so undecided about the future I had a kind of neurotic illness that is really a way of preventing living by not living. And I also found that the most powerful killer of love in anxiety. Thank God Helen stood by me.'

John looks at him directly.

'If you are unsure about certain issues get advice from anyone whose views you trust – perhaps that's the reason you're here today, and then make up your mind on what you think is best. Have your opinions ready. Also there may be opposition from some unknown source, and propaganda could be the main weapon. It can be a dangerous one for anyone to use because if one holds it long enough it can move about like a snake and strike the originator rather than the intended victim. Also don't ever lose your temper.'

'Of course. That would be a sign of instability in a clinician.'

'Good point, Graham,' John replies slapping his hand on his knee for emphasis. 'That's exactly what some panellists try to find out in different ways. So watch out for being set-up on the clinical angle of things.'

'Yes. Is he dependable in an emergency or crisis?'

'Precisely,' John replies and pauses for a moment. 'When you've been treating very ill patients for a long time there are certain things that medical textbooks cannot teach you. I suppose a useful word is experience but even that doesn't tell the complete story. You should become familiar with each patient's needs and almost live through them yourself before you can properly practise. I'm not talking about morals here, Graham; I'm talking about facing the truth together with each patient and trying your best to give them peace of mind. In order to do so you have to develop a sense of compassion and even serenity inside you — because, sometimes truth can be a terrible weapon of aggression. So be careful with it.'

'I see – the difference between a good and bad clinician.'

'Yes. I'm sorry if you think I'm giving you a lecture and stuff you already know, but keep it foremost in mind during the interview. Weakness and indecisiveness show through very quickly'

'Perhaps more so behind an interview desk than in the wards.'

'Yes. You're not let get away with much during modern interview tactics, especially if there is a good chairman.'

Graham reflects for a moment on all that John has been saying.

'You're right, Dr. Nicholson. I think in the long term it is probably better to burn out than to fade out.'

'Well said, Graham. That shows determination,' John replies with a sigh. 'As I grow older thinking becomes more difficult, yet the heart more eager, courage more reckless as time – that consumer of everything -- gets shorter. As long as I have a need I'll have a will to live. Satisfaction can be disastrous.'

'You must always want to know what's around the next corner.' Graham says.

'Yes. Science may have made great advances in certain areas but it has yet to find a cure for human apathy. One last thing, select your friends with great care. Your opponents will soon choose you.'

*

At the end of a long day in the laboratory John Nicholson looks at the computer screen in front of him. Even positive advances can extract a price in negative terms -- he is totally exhausted. In spite of this his eyes are ablaze. He reads the result for the third time and is almost moved to tears. He's alone in the vast room so he walks around it several times and then back to the computer. He reads it again. Yes, these are the results he has been looking for, and they vindicate everything he's been desperately searching for. Now no one can dispute the science, the truth and the sheer logic of it all. No one can deny its worth. It appears in front of him like an absolution. If he had failed in this project – and some failures are written large on the calendars of past years – he would never be forgiven and probably worse. Now this is the answer to a problem that has plagued him for a long time. Now he can hold his head high and defend his work.

He is totally convinced it is reliable, dependable and safe and has major advantages. Double blind clinical trials using his method and conventional agents show his results are superior, more flexible and effective. As a result he will be prepared to defend the results anywhere.

He works hard at preparing publications for the medical press, which are assiduously peer reviewed. His team makes a remarkable impression on the scientific world.

The effort imposes a strain on his relationship with Karita and marriage arrangements have to be postponed. Also she still travels with Nolan to locations in Ireland and Scotland. The sting is she refused to come to Florence, but she has time for location work. No problem there!

She protests vigorously. 'But John, that's different. That's my whole work at present and everything has to be just right. I've told you many times how essential it is to meet deadlines that one has promised. Dear John you must understand what I'm going through.'

'I'll try, Karita. I promise.'

'It's going to be different when it's all over. You'll see John – completely different. We'll have more time for each other. And of course you're working sixteen hours a day trying to prove your research deserves a place in the scientific world. I realise how hard it is to convince so many vested interests your way could have great advantages – if only you were given a proper chance.'

'Yes, Karita. You're right. I need more help and staff to make a greater impact.'

Months go by and work proceeds in both camps with polite exchanges, hiding subtle innuendoes. Then news arrives for John in the form of anonymous donations from three wealthy sources. They wish to see his research expand and for him to be able to explain to the scientific community the complete validity of his work and its trustworthiness, which could even make animal testing redundant.

An enormous challenge.

# 10

The University of Cambridge welcomes the financial backing for Dr. Nicholson and a lot of legal documentation and trusts have to be set up before money can be released. It is a major boost for the research effort. It means Dr. Nicholson can gather around him the best brains in information technology and biometrics including Dr. Graham Sinclair to carry out trials on cancer patients to a much greater extent.

This takes time but the machine moves on inexorably and becomes bigger and better; publications increase enormously and considerable interest is generated abroad. Cambridge has achieved an international reputation with in vitro testing and complicated drug manipulation. Protocols are created that begin to rock the foundations of the establishment – worldwide.

In recognition of John Nicholson's rapid and outstanding pioneering work the Central Administration Board of the university -- encouraged by Sir Kenneth's remark that genius does what is imperative, and talent does what it can -- recommends that Dr. John Nicholson be offered a new Chair in Oncology and Experimental Pathology and conferment of title Professor John Nicholson would be acceptable to all from now on. John is thrilled and a special reception is arranged. This highest accolade the university could give him is a tremendous boost in his efforts to convince the scientific community that cancer research and treatment should change -- if they could only open their eyes and minds the future is already here. Old fashioned thinking and teaching should be gradually phased out as these newer, more immediate, and scientifically provable protocols genuinely do work and work well.

When Karita hears the news she goes wild with excitement. She can hardly believe it but on reflection feels a little foolish not seeing it coming. Ever since John came back from Florence – which he would not talk about – he seems like a changed man, more ruthless, more impatient, less trusting, almost a barrier to straight forward communication. He answers simple questions that only require a 'yes' or 'no' but usually with another question 'why?' What has changed him in a short period of ten days? She wishes she'd time to go with him. She must remember in future.

In spite of his academic elevation John cannot neglect his personal life. He's already complained that Karita is spending a lot of time with Nolan. Is this necessary? One evening the three of them meet in his cottage outside Cambridge. Things are uncomfortable and electricity is in the air. The publishers of The Trilogy are thrilled with Nolan's illustrations. They would definitely enhance the overall appeal of the work instead of a group of photographs.

Months pass again. The Trilogy is published in London and soon after a US film company spots the potential for the screen based on the work.

The controversy between animal testing and in vitro testing intensifies.

The battle is on.

# 11

The Royal Swedish Academy of Sciences normally starts a series of meetings in the beautiful city of Stockholm to determine one of their most important functions – The Prize. Various selection committees consist of senior Swedish scientists who have gathered information and reports from leading academics throughout the world. Members of each committee – usually five in number -- are as distinguished as the potential Laureate and the selection of winners is an unbiased task similar to the stringent processes that surround such events as the Olympic Games.

The prizes to be awarded were chosen by Alfred Nobel in his famous will of 1894 when he created prizes in physics, chemistry, medicine, literature and peace. The prize in medicine is awarded by the Karolinska Institutet located in Solna near Stockholm; the economics prize came later. It has now become for scientists and laymen alike the ultimate symbol of excellence and is intended to make prize holders life scholars in the eyes of the world.

There are certain stipulations laid down such as candidates must be living and no more than three can share a prize. It is not intended for outstanding individuals nor is it a lifetime achievement award. It is given for specific scientific achievements, inventions and discoveries. The election process involves massive amounts of paper work. Eminent academics throughout the world are invited to select and report on anyone they consider worthy of consideration.

One day Professor John Nicholson receives a phone call from a Professor McNeil requesting a meeting in Cambridge at John's convenience. No details are given but they agree to have lunch in early May. Professor McNeil arrives on the day appointed and John meets him at Cambridge rail station. They drive to the hotel where John has booked a table overlooking the River Cam. The location and day are perfect.

'Nice to meet you, Professor Nicholson,' McNeil says at the station. There is only small talk on the way to the hotel. John is puzzled. What does this distinguished professor from London want to discuss and apparently in relative secrecy?

'Please be seated, Professor. I hope the view is to your liking.'

'Beautiful, although I've been to Cambridge many times and at each visit it seems to become more appealing.'

'I know what you mean. It has a certain magic.'

A waiter appears with menus.

'Professor Nicholson, may I call you John?'

'By all means. I would be delighted.'

'Mine's Alex.'

'Now we can relax a little and study what is in front of us first.'

John makes his choice quickly but Alex takes more time and looks around the half-empty room several times assessing if anyone is within earshot.

'John. I've admired your work for sometime now and your publications are outstanding.'

'Thank you. A lot of thought went into them and, so far, things have worked out satisfactorily.'

'You're intending to continue working in the same field?'

'Yes. We've now got a larger team in the department that allows us to get results more rapidly – thanks to those wonderful benefactors who wish to remain anonymous.'

'I see. So more of the same quality research to come?'

'Yes. I hope so. That seems to be the way things are going.'

Alex plays around with his knife and fork apparently thinking of what to say next.

'John. You may know I'm in charge of Oncology at the Hammersmith in London. Have been now for nearly ten years and ...'

'Of course. Your work is well known and admired.'

'Thank you, John. But that's not the reason for my visit. However, before we come to that may I put a particular problem to you in Oncology and see how you would solve it. I had a pleasant hour on the train to formulate a hypothetical situation and I'd like to see how you'd approach it and possibly solve it.'

'Goodness me...' John smiled. 'You're going to give me a verbal jig-saw puzzle and then see how long I'll take to solve it.'

'Well yes – but more like a game of chess.'

'OK then. Fire ahead.' John leans forward but is interrupted by the waiter who wishes to take their orders. When completed Alex starts to outline a hypothetical case scenario with multiple questions. When finished he opens a hand. 'Over to you.'

John produces a pencil and starts to scribble on a paper napkin. Diagrams and arrows rapidly appear and grow larger in rapid succession. A second napkin is required and many chemical equations appear. He closes his eyes for a moment and then slams the pencil down on the table.

'I've got it. You should proceed thus.'

He then explains in detail and in rapid succession each step in solving the problem. Alex sits back and gives a sigh of amazement.

'You know, John. It took me and my colleagues nearly two days to sort that one out and here you solve it before the soup arrives. Extraordinary! What intrigues me about your approach is the originality of thinking. No one has ever been able to see this line of inquiry as clearly as you.'

'Well Alex. Now that we have computers as never before,

which can produce results within seconds that used to take months to work out, progress can be rapid.'

'True, true. A unique advantage. But still the computer must have someone to tell it what to do. It just doesn't work in a vacuum.'

'Of course. It needs the right information fed into it and also needs to be kept on the straight and narrow.'

Alex smiles. 'Straight and narrow,' he mumbles. 'And what a difficult problem this could be for some people.'

'Yes. Constant vigilance is necessary and corrections need to be made all the time.'

The rest of the meal is more relaxed with talk about science, government policies, The NHS and university problems. After coffee Alex asks John to join him for a short walk along the river bank.

'John, what I've really come to see you about is this. I have been approached by the Royal Swedish Academy of Sciences to ask if you would be prepared to accept the Nobel Prize in Medicine, that is, if the selection committee fully approves. In other words would you let your name go forward for selection?'

John stops walking, looks straight at his visitor's face and says.

'Of course, I would. But surely this is a secret process and names are not published until the last moment.'

'Correct. But approaches are made all over the world and I've been asked to come and talk to you and also ask you to keep our conversation confidential for the moment.'

'Why, yes. It must be a very stringent process.'

'The RSA insists on keeping the Olympian aura that envelops the whole process. It is one of the most secretive societies imaginable and all the reports, recommendations, inquiries and discussions – especially the voting system – are among the most guarded secrets in the world. The statutes laid down go something like this.'

He takes out a sheet of paper from his inside pocket and reads.

"Proposals received for the award of a prize, and investigations and opinions concerning the award of a prize may not be divulged. Should divergent opinions have been expressed in connection with the decision of the prize-winning body concerning the award of the prize, these may not be included in the record or otherwise divulged. A prize-winning body may, however, after consideration in each individual case, permit access to material which formed the basis of evaluation and decision concerning a prize, for purpose of historical research. Such permission may not be granted until at least 50 years have elapsed after the date on which the decision in question was taken." There have been occasional breaks but generally speaking the whole process works well.'

Alex looks at his watch and says.

'I'd better make my way back to the station. I wish to catch the 4.10 to London for an important meeting this evening.'

'Don't worry Alex. I'll drive you back to the station in good time.'

*

Regular meetings continue to take place in The Royal Swedish Academy of Sciences. The selection process is a complex procedure. After Professor McNeil sends in his report the committee asks him to attend a couple of meetings and is pledged to complete secrecy. As McNeil enters the panelled room he recognises most of the Swedish elite. He feels privileged to be given the opportunity of helping to decide a significant international issue. He explains the essential features of Nicholson's and other candidates' work he is asked to nominate. He has placed Nicholson at the top of his list.

In early October the selection committee has come to its final decision but still members are sworn to secrecy for another while.

Weeks later the Academy calls a press conference. The Nobel Prize conference has been called for 11.30 in the morning. This impressive event is always held after the final ceremonial vote and always starts on time. Announcements are made and the news is flashed around the world in minutes about the new Nobel Laureates.

In the same afternoon there is a champagne reception in the fine Assembly Hall. McNeil and other colleagues gather together to discuss the results. Although John Nicholson is not present Alex McNeil has been able to phone Sir Kenneth Richardson to break the good news and leaves it to him to tell John Nicholson.

*

John Nicholson arrives in Stockholm a few days before the official ceremony on the 10th of December. Karita remains in London for a couple of days for book signings and receptions and then travels home.

Most people would agree that time heals and one of the benefits of age is that the truce between the past and present is easier to accept. This is generally the advice given to the afflicted. Some will even say that the ability to forget can be a blessing. Time reduces the effects of grieving and removes the edge of guilt – it devours everything. John had tried to bury the memory of Linda Lindstrom, his first wife, away and not discuss it with anyone – even with Karita – and he had remained resolute.

But now, alone in his hotel bedroom, he looks over his past and is overcome by a strange and cold fear. He had brought Linda's diary with him and as he touches it there are jumbled feelings of pleasant love and passionate encounters, but also guilts, sufferings and betrayals.

Time suddenly ceases; space is suspended around him. Everyone of his senses is alerted by a swift surge of memories; the luxuriant blonde hair flowing back in waves over the ears and halfway down the neck, the exquisite gold earrings dangling provocatively from small chains and the width of brow and tilt of chin suggesting a wild energy and reckless nature. And those sparkling light blue eyes were unforgettable. Memories of their first meeting in Stockholm seem as if it were yesterday – like the pages of a symphony, seen in silence, yet heard in glorious harmonies through the caverns of his mind. Even after all these years her music had the power to overwhelm and her smile was like a summer's day.

He touches the diary again almost afraid to open it. Special things, such as Linda's diary, have an uncanny power to remind us that our past is very real. Yet long practice in medicine had made him a sceptic, a cynic, making it hard to be emotionally involved; it was the only way to survive the stresses and strains of daily life. In spite of his Celtic background he does not accept the concept of ghosts, yet deep down in his subconscious is an area haunted by unexorcised memories, problems and predicaments. Small sorrows are easy to talk about; great ones remain silent.

He sits for a while in a state of confusion and paralysis unable to unpack.

Then suddenly his own company becomes intolerable. He quickly puts on his warm overcoat and heads for a familiar bistro in Gamla Stan where the owner serves drinks on a marble counter and his wife provides delicious a la carte choices. In the corner sits a young girl playing soft slow music on a piano. Here he is known and welcomed. He selects a quiet position and feels not too alone. It was here he spent his first evening with Linda Lindstrom who is now no more – but he can still hear her voice. It is soft, good and kind.

*

The big day arrives.

John is briefed on procedures and protocols of the ceremony itself and it is hoped he will be able to cope with all the pomp and grandeur of the occasion – full dress and decorations are to be worn. The finest memory of John is the much-feared audience with the King of Sweden. By tradition the King would allow four to five minutes alone with each Laureate. When John's turn comes he nearly trips over the carpet at the entrance to the Royal apartments. Five minutes pass. Seven minutes pass. Then ten minutes. Unusual audience for one single candidate. John comes out of the private chambers and everyone wants to know what they talked about. Apparently it was a polite but slightly heated discussion about the pros and cons of hunting.

The major ceremony takes place on the 10th of December each year at the Stockholm Concert Hall. Organised by the Academy's Nobel Foundation the award ceremony is the climax of a very busy week. The Laureates take centre stage in the Concert Hall where each receives the Nobel Prize Gold Medal engraved with the portrait of Alfred Nobel, a personal diploma and a cash award from the King of Sweden. In Oslo, the Nobel Peace Prize Laureates receive their prize from the Chairman of the Norwegian Nobel Committee in the presence of the King of Norway.

Later he is to discover that among the sponsors for his nomination were Sir Kenneth Richardson, Dean of Medicine at the University of Cambridge, Sir Neville Chance of the Tate Gallery in London, Professor James Kennedy of Trinity College Dublin and Professor Gustav Isselherg of the Helmuth Institute in Stockholm.

So John Nicholson's journey to Oslo and Sweden is for a double celebration – the conferring ceremonies and the marriage to beloved Karita.

An important part of the Nobel Prize Awards is the presentation of the Nobel lectures by the Laureates. In Stockholm, the lectures are given days before the Nobel Prize Award Ceremony. In Oslo, the Nobel Laureates deliver their lectures during the Nobel Peace Prize Award Ceremony.

Professor John Nicholson feels obliged to add several personal items at the end of his address:

'Your Majesties, Your Royal Highnesses, Distinguished guests, Ladies and Gentlemen I come here today to accept this great honour with pride and humility – pride because to quote Newton "If I have seen further it is by standing on the shoulders of giants" and humility because there are many others who have shaped my life and work.'

'Not least my parents; mother always insisted that there are two choices open to us – the right one and the wrong one. The outcome will always depend on you. My father was a man of principles. They were not normally given in words but by example. To both I remain eternally grateful. I also had teachers who instilled a sense of dedication, of morality and of idealism that once taught could never be untaught. The conviction of not just following or imitating others gradually developed but of thinking for oneself.

An experiment is designed to make Nature speak lucidly and logically. After that we only have to listen. In science there are no new facts, only those that have been seen by others who have observed them without noticing. We pass this way but once and what ever purpose or reason we hold dear must be pursued with all the vigour and might we have and especially in times of adversity. Like others I've had my share; it can be easy to give in and give up and many have done so. Others have held out against some of the greatest odds, which have been planned to destroy them and their work and have overcome them with dignity.'

'However, there can be no gain without pain and without strong support from friends and colleagues the outcome could have been a triumph for those who wish to destroy. I therefore now acknowledge those who have come to my aid especially in the darkest hour. Fortunately time is what stops everything happening at once. I have found in my research that time pushes the horizon further away, to recede, which makes life frustrating. Also, as one's endurance also lessens the importance of the work becomes more intense and there's always the feeling of efforts being unfinished.'

'I have, unfortunately, noticed an unwelcome gene in academic life. We are now creating a new generation of scientists, especially in the field of bioscientists where a change from the laudable advice to "publish or perish" to "patent and profit."

'A lot has yet to be done. Future empires will probably be of the mind. It is one of my hopes to hand on the torch to other like-minded people to continue the work that all living creatures desperately need in the years to come. I have learned that out of adversity greater and more fundamental dedication can come forth if the lessons of such people are adhered to – and we all choose the right thing.'

'Thank you for bestowing on me one of the greatest honours available. It vindicates the toil, sweat and tears, the suffering of patients and fellow creatures who have in some way contributed to this particular occasion and in that I am certainly not alone.'

'A medal glitters but it can also cast a shadow. So I say:

To those I have damaged I beg forgiveness,

To those who would harm me I entrust to our Creator,

To those who are oblivious of me, whether intended or not, you do well,

To those who will remember me I also will remember you with heartfelt thanks.'

And even to this day the battle goes on to find the best method of evaluating chemotherapeutic agents.

Lightning Source UK Ltd.
Milton Keynes UK
03 April 2010

152281UK00002B/28/P